THE WHITE HOUSE

WASHINGTON

National Environmental Technology Strategy

The quarter century since 1970 stands out as a period of great environmental and economic achievement, and all Americans can look back on our nation's accomplishments with pride. Yet our work has just begun. We still face major challenges both domestically and globally if we are to achieve continued economic growth and a healthy environment.

Today we have a special opportunity to reflect on the past and define a vision for our future. Our views about the environment have evolved and become more sophisticated over the past 25 years. One lesson we have learned is that economic growth and environmental stewardship go hand in hand. A clean environment means a higher quality of life, and technological advancement means economic growth and better jobs for American workers.

For the past two years our Administration has sought the views of Congress, the states, communities, industry, academia, nongovernmental organizations, and interested citizens on ways to spur the development and use of a new generation of environmental technologies. This document represents the views of thousands of individuals who participated in events around the country to help us craft a national environmental technology strategy that will put us on the path to sustainable development. We thank everyone who participated in this endeavor.

If our environmental technology industry is to remain competitive in the global marketplace, we must implement actions today that will be responsive to tomorrow's problems and needs. Meeting future challenges will require our regulatory system to adapt to a changing world by promoting the innovation that will ensure protection of our environment in a cost-effective manner. We want a government that offers opportunity, rewards innovation, and demands responsibility. And we as individuals must develop a more sophisticated understanding of the linkages between our environment and the economy.

Now the time has come for creative action and bold steps. Let us pledge to use technologies wisely for they are the bridge to a sustainable future. Our foresight will define the structure of that bridge. Our creativity will allow us to build it. And our commitment will determine how quickly we cross it.

Bill Clinton   Al Gore

This document was prepared under the guidance of the National Science and Technology Council (NSTC). The NSTC, chaired by the President, is a cabinet-level council charged with coordinating science, space, and technology policies throughout the federal government. An important objective of the NSTC is to establish clear national goals for federal science and technology investments. The NSTC includes the Vice President, the Assistant to the President for Science and Technology, the Cabinet Secretaries and agency heads with responsibility for significant science and technology programs, and other key White House officials.

This national strategy builds on the ideas and recommendations in the earlier NSTC document, *Technology for a Sustainable Future*, released in July, 1994. Copies of the first report are available through the Interagency Environmental Technologies Office, at the address listed below, through the Internet (etstrategy@gnet.org), or by calling 1-800-ENV-6676.

April 1995

The National Science and Technology Council is actively seeking suggestions for improving federal policies related to advancing environmental technologies. We encourage you to contact the offices listed in the appendix of this report or send your ideas to:

Executive Director
Interagency Environmental Technologies Office
955 L'Enfant Plaza North, SW, Suite 5322
Washington, DC 20024
Internet: etstrategy@gnet.org

Printed on recycled paper.

Bridge to A Sustainable Future

National Environmental Technology Strategy

Bridge to a Sustainable Future

TABLE OF CONTENTS

EXECUTIVE SUMMARY

As a nation, we seek long-term economic growth that creates jobs while improving and sustaining the environment. Achieving these goals requires an environmental technology strategy that addresses the need to remediate past environmental damage while helping us shift from waste management to pollution prevention and more efficient use of valuable resources.

In the 25 years since the first Earth Day, our nation has made considerable progress in responding to threats to public health and the environment. Yet major challenges remain, and the decisions we make today will define the path we travel and the quality of our children's lives on Earth Day 2020. We can put in place today a set of policies and programs that will establish a new course for the development and use of environmental technologies into the next century. We must, however, broaden our environmental tool kit, replacing those instruments that are no longer effective with a new set of tools designed to meet today's challenges and tomorrow's needs.

Over the past six months, thousands of stakeholders have discussed the merits of setting a new course and identified the tools we will need to change our present direction. From these discussions, a core set of five themes emerged that can be used to guide future activities. These themes are designed to establish the framework for partnerships, goal setting, policy development, and action. Within each theme we present a series of findings, goals, and initiatives that together articulate a technological path that will lead us toward sustainable development. These are summarized below.

Performance, Flexibility, and Accountability

Together, we need to develop a new generation of incentive-based policies and programs that stress performance, flexibility, and accountability. Our focus must shift to systemic whole-facility, multi-media, and multi-sectoral approaches. These programs need to be credible and inclusive, allowing the participation of all interested parties in the development and implementation of new technological solutions and the information and organizational structures that support them.

Finding: Although federal and state environmental regulations define and shape the market for environmental technologies, their inflexibility and unpredictability in some cases limit the development, commercialization, and application of new environmental solutions.

Goal: Achieve continuous improvement of the environmental performance of U.S. industries, using the most advanced technologies and cost-effective means possible, by strengthening incentives for innovation within the regulatory system.

Initiatives: The Environmental Protection Agency will work with the states, industry, and environmental groups to advance several initiatives that will spur the development and application of environmental technologies. These include a new, more flexible approach to environmental management under Project XL (Excellence and Leadership); an evaluation of the effectiveness of third-party auditing of the environmental performance of industry; an ongoing Environmental Technology Initiative directed to reducing regulatory and market barriers; and an effort to streamline permitting and approval programs.

Innovation for Environmental Results

Over the long term, research, development, and demonstration programs should facilitate a broadening of focus from reacting to environmental damage to anticipating and avoiding it, and producing more while using fewer natural resources and less energy. As our focus expands, however, we must ensure that pressing remediation problems are dealt with and cost-effective compliance with existing regulations is achieved. Government programs need to facilitate and accelerate private sector research, development, and commercialization efforts.

Finding: The federal government plays an important role in funding basic and applied research and development (R&D) that is key to the development of future generations of environmentally critical technologies. The federal government also facilitates private sector and cooperative investments in needed R&D, by reducing uncertainties caused by the regulatory, verification, and permitting systems.

Goal: Increase the overall productivity of our nation's energy, food, manufacturing, transportation, construction, and service sectors through environmental technologies and practices that significantly reduce the use of energy, materials, and other inputs. Environmental technologies are needed to improve understanding of atmospheric, terrestrial, and aquatic systems and to remediate and restore environmental damage in a cost-effective manner.

Initiative: Within the next six months, the federal government, together with the private sector and state and local governments, will update research, development, and demonstration priorities for environmental technologies.

Finding: New approaches are needed to demonstrate innovative environmental technologies in order to answer key questions of economic viability and acceptance, expedite commercialization, and foster acceptance by stakeholders and regulators.

Goal: Accelerate and facilitate the demonstration of promising environmental technologies while reducing costs.

Commercialization

We must bring together innovations with the necessary capital and market information — and do it in a timely manner, especially for small businesses. To accomplish this objective, we will need a wide range of complementary policies to support investment and the diffusion of successful technologies. We also need to commit to fostering environmental awareness and institutional and technological capacity in other parts of the world to encourage the use of appropriate environmental technologies.

Finding: The lack of a system that credibly verifies the performance of innovative environmental technologies is a significant factor in limiting promising environmental solutions.

Goal: Develop, through the federal government and the states working together, credible performance information for precommercial environmental technologies.

Initiatives: The federal government will work with the private sector to establish a market-based verification process for environmental technologies. This process will be available nationally for environmental technologies within three years. In addition, the federal government recently launched the Rapid Commercialization Initiative (RCI), which is intended to accelerate the commercialization of near-commercial environmental technologies. Over the coming year, ten technologies will be commercialized through this new program.

Finding: Financial uncertainty and a high level of risk limit the availability of investment capital for environmental technologies.

Goal: Ensure that adequate investment capital is available for the development, commercialization, and use of environmental technologies, both for entry into the domestic market and to support an active export industry.

Finding: Due to tremendous domestic demand and the limited financial strength and export experience of small- and medium-sized companies, the U.S. environmental industry has mostly focused on the domestic market. Many U.S. environmental companies have difficulty competing internationally, resulting in missed opportunities to support or create new U.S. jobs.

Goal: Increase U.S. environmental technologies exports to support and create new, high-paying U.S. jobs and to contribute to the achievement of sustainable development.

Finding: Sustainable development and the dissemination of environmentally beneficial technologies are hindered by inadequate infrastructure, human and institutional capacity, and policies and regulations in developing countries. Investments are critically needed to create an international environment that enables the diffusion of environmental technology.

Goal: Build a foundation for environmental stewardship and sustainable development internationally by implementing a coordinated set of activities to facilitate avoidance of environmental harm and remediation through the development, adaptation, and use of environmental technologies.

Initiative: The U.S. Agency for International Development, as part of its Initiative for Environmental Technologies, will work in partnership with the private sector to focus development assistance on addressing critical environmental challenges in developing countries.

Sustainable Communities

Communities, both urban and rural, will need to make significant advances toward sustainability over the coming decades, using environmental technologies appropriately and wisely. New generations of more efficient and integrated infrastructures for energy, transportation, water and wastewater treatment, and communications must be carefully designed today to ensure a more sustainable technological base for tomorrow's cities, towns, and neighborhoods.

Finding: Our nation's future strength will in large part be built on the viability of our communities. We must make choices today that increase the sustainability and desirability of our cities, towns, and rural areas if we are to preserve our natural environment and build a strong domestic economy.

Goal: Develop and implement sustainability plans in many U.S. communities and make significant progress toward achieving sustainable communities over the next 25 years, increasing the quality of urban, suburban, and rural life, and reducing our use of energy and natural resources.

Initiative: Communities will better integrate the development and use of environmental technologies into sustainability plans and activities.

Learning and Working Together

Achieving our goals will require more effective, open, and productive collaboration among stakeholders. It will require new models of gathering, sharing, and analyzing information; coordinating work; and educating and training professionals, policy-makers, and the public.

Finding: Domestic and global markets for environmental technologies are influenced by actions of governments and the private sector. An improved relationship among all levels of government and the private sector can stimulate investment in environmental technologies and increase their diffusion.

Goal: Ensure that the federal government becomes a more accessible partner with the private sector in advancing the development of environmental technologies.

Initiative: The federal Interagency Environmental Technologies Office will act as an information clearinghouse and bridge between the federal government and the private sector in advancing environmental technologies.

Finding: Education and training programs at all levels, from elementary schools to universities, have not sufficiently integrated information on the challenges and opportunities associated with sustainable development and the role of environmental technologies in achieving it.

Goal: Build an integrated, interdisciplinary environmental education and training system for students at all levels over the next decade.

Initiatives: Several initiatives will be undertaken to advance environmental education and training including a program to foster and promote a new environmental ethic within the system of higher education; an effort to work with the, industry, the states, municipalities, nongovernmental organizations, and the educational community to develop a blueprint for building successful partnerships to support environmental education and training; and a program to build on the nation's network of community colleges to advance programs related to sustainable development and environmental technologies.

Finding: The considerable national investment in environmental monitoring, assessment, and related information systems made by all levels of government and the private sector should be fully and cost-effectively linked to support efforts to achieve sustainable development.

Goal: Improve the nation's environmental monitoring data and information systems substantially over the next five years through public-private partnerships designed to maximize opportunities for developing and sharing information essential for achieving sustainable development.

Initiative: Through the INSIGHT 2000 initiative, the federal government will establish or broaden a series of partnerships to improve access to federally sponsored environmental data. Systems will be established for disseminating and applying environmental information to enable effective environmentally related decisionmaking by businesses, regulators, and the public.

Moving Forward

Moving forward effectively will mean moving forward together. Achieving our vision of sustainability will require that individuals and organizations recognize their unique ability to contribute to the goals set forth in this national environmental technology strategy. Industry, labor, communities, nongovernmental organizations, individuals, the states and federal government, and nations around the world all have important responsibilities in this effort to advance environmental technologies. The key to progress is building on the strengths of each sector in order to achieve the goals collectively that cannot be achieved individually.

Part I

A New Challenge

Chapter 1

A New Challenge

Earth Day 1995 is a time to celebrate, reflect, and commit. Much has been accomplished over the past 25 years, and much remains to be done. Today, it is within our power as a nation to have a thriving economy as well as a healthy environment. However, attaining these goals will require a strategic transformation of environmental and economic policy. To accomplish this, we have listened to the advice of individuals throughout the country in developing this national environmental technology strategy.

What is our vision of the future? As a nation, we seek long-term economic growth that creates jobs while improving and sustaining the environment. Achieving these goals requires an environmental technology strategy that addresses the need to remediate past environmental damage while helping industry shift from waste management to pollution prevention, efficient resource use, and industrial ecology. A forward-looking approach will help companies become more competitive by lowering their energy and resource needs while reducing or eliminating their waste cleanup and disposal costs. Nationally, it will create economic growth by capturing the rapidly growing market for clean technologies and shifting money from consumption of resources to investment in new plants and equipment. Globally, it will help developing countries more rapidly utilize sustainable technologies in their industrial and service sectors. Being the world leader in clean technologies and processes will make our economy more competitive, increase our exports, and help us build sustainable communities for a more sustainable world.

To achieve this vision, we will need a new approach that enables us to harness the transforming powers of science and technology. We need a regulatory system that stresses high environmental standards and encourages flexibility and creativity in meeting those standards. We must reinvent the way government carries out its responsibilities. We as individuals need to reflect on our own behavior as consumers. Perhaps most importantly, we all need to work in partnership to accomplish together what cannot be accomplished alone.

Learning from the Past

This call for a new approach to meeting the challenges facing the nation arises from a recognition of the limitations of past policies. Over the last 25 years, environmental legislation and regulatory policies have achieved impressive victories. But the capacity to foster economic and environmental progress simultaneously through traditional regulatory-based, command-and-control approaches has begun to diminish.

Regulations have channeled massive investments into certain types of technologies at the expense of other equally effective approaches. Though many technologies have enabled us to achieve environmental goals, they have not always done so in the most cost-effective manner. Worse yet, our regulatory system has fostered a particularly adversarial climate among industry, government, and nongovernmental groups. This antagonistic climate has undermined trust and reduced the opportunities for experimentation that are key to technological and social innovation.

Today, we face new classes of environmental problems that are more diffuse than those of the past and demand different approaches. Since Earth Day 1970, the vast majority of the most significant "point" sources of water pollution — large industrial facilities and municipal sewage systems — that once spewed untreated wastes into rivers and lakes have been controlled. The most important remaining sources of water pollution are diffuse and widespread: sediment, pesticides, and fertilizers that run off farmland, oil and toxic heavy metals that wash off city streets and highways. Pollution from these sources cannot be controlled with the same regulatory techniques employed to check emissions from large industries or sewage treatment plants. Further, problems of global scale — biodiversity loss, ozone depletion, and climate change — have emerged. These problems will require innovative global solutions. We are also recognizing that controlling emissions alone, no matter how successful, will not achieve an environmentally sustainable economy.

Defining Environmental Technology

An environmental technology is a technology that reduces human and ecological risks, enhances cost effectiveness, improves process efficiency, and creates products and processes that are environmentally beneficial or benign. The word "technology" is intended to include hardware, software, systems, and services. Categories of environmental technologies include those that avoid environmental harm, control existing problems, remediate or restore past damage, and monitor and assess the state of the environment.

Technology has been called the engine of economic growth. It is responsible for as much as two-thirds of the increase in our nation's productivity since the Depression. But technology is not a panacea for the formidable environmental challenges that face our nation and the world. It is a tool that is applied to achieve an end. If the end is sustainable development, then our individual and collective actions will need to reflect a commitment to applying technology appropriately. If we are smarter about the technological choices that we make, we can provide the quality of life that we all desire for ourselves and our children.

The Beginning

The process of developing this national environmental technology strategy began nearly two years ago. The first phase culminated in July 1994 when Vice President Gore released the National Science and Technology Council report, *Technology for a Sustainable Future*. This report initiated a national dialogue that led to the strategy contained in this document. The National Science and Technology Council conducted a series of technical workshops, policy symposia, and a White House Conference to determine how best to foster environmental technologies. More than 10,000 individuals attended some 30 events held throughout the nation (see appendix).

In effect, this strategy was developed in Seattle, Washington; Burlington, Vermont; Atlanta, Georgia; Kansas City, Missouri; El Paso, Texas; and in dozens of other cities across the United States. In those meetings, the discussion included individuals from large companies, small businesses, organized labor, farms and ranches, federal agencies, universities, community colleges, environmental groups, banks and investment firms, and state and local governments. The contributions of these individuals were responsible for developing the ideas contained in this document. These same individuals, and many others, have the capacity to turn this strategy into a reality.

Defining the Future

The outlines of the road ahead can already be discerned in the path we have traversed. We are progressing from an environmental paradigm based on cleanup and control to one including assessment, anticipation, and avoidance. By the start of the next century, intensified expenditures to develop technologies that avoid environmental harm will begin to pay off. Agricultural practices will be less wasteful and more sustainable, manufacturing processes will be more efficient in the use of resources, consumer products will be designed with the environment in mind, and the infrastructures that supply energy, transportation services, and water supplies will be more resource efficient and environmentally benign. Investments in cost-effective remediation will have cleaned up a large portion of existing hazardous waste sites. Our ability to respond to emerging problems will be aided by more advanced monitoring systems, microsensors, and data analysis tools that continually assess the state of the environment locally, regionally, and globally. Finally, we will have developed effective ways of restoring or recreating severely damaged ecosystems to improve the long-term health and productivity of our natural resource base.

> ***"The best way to predict the future is to invent it."***
>
> *Alan Kay*
> *Apple Computer*

Moving our economy and society onto this new path is possible if we are willing to reorient our policies, develop new strategies, and closely coordinate the actions of the public and private sectors toward a set of shared, long-term goals. This national environmental technology strategy is intended to achieve that end. It is designed to serve three purposes. The first is to articulate a vision of the future and the role environmental technology will play in shaping that future. This vision must be built on an understanding of the strengths and weaknesses of past policies and actions. The second is to develop a strategy that helps define the roles of the many individuals and organizations who are needed to implement the strategy. The third is to chart a course by offering suggestions for strategic goals for all partners in this important national endeavor.

The strategy outlined in this document is national in scope. However, we recognize that the causes and solutions of environmental problems transcend national boundaries. In the post-Cold War world, our foreign policy is becoming defined in terms of economic competitiveness and environmentally sustainable development. Increasingly, we must address transnational issues of global significance, such as population growth, public health, adequate food and clean water, climate change, and other environmental concerns. We must apply our energies to help other nations develop the institutional capacity to address these problems and achieve their own environmental and economic goals. This will not only build stronger and more stable foreign economies but also stronger foreign markets for U.S. environmental technologies.

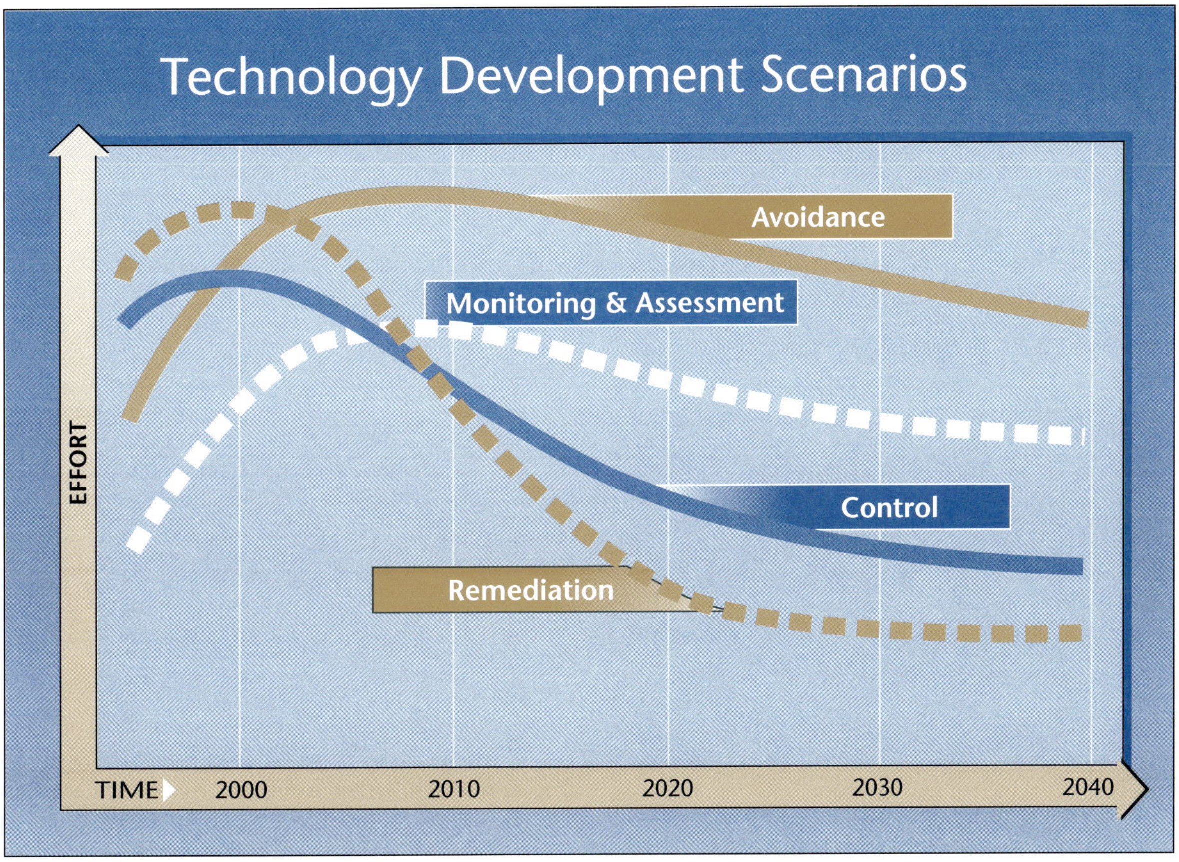

Our efforts to develop environmental technologies will change over time as we place greater emphasis on pollution prevention approaches with high efficiencies in resource use and on the restoration of critical ecosystems. At some point in the future, we will have moved from an environmental paradigm built on control and cleanup to one also based on assessment, anticipation, and avoidance.

Although the federal government is a catalyst for action, it is only one player among many. Accordingly, this strategy is a call to action for industry, academia, government at all levels, nongovernmental organizations, and interested members of the public. Throughout this document, highlights of ongoing initiatives by various organizations are presented to illustrate key components of the national environmental technology strategy.

In parallel with the process that led to this strategy, specific action plans for implementation are being developed by the federal government as well as other partners. These plans are dynamic and will continue to evolve as communities, businesses, states, and the nation travel the path toward a sustainable future.

"I do not believe it was a coincidence that the first Earth Day in 1970 immediately followed our first mission to the moon. This new awareness, the new ability to see the earth floating in space, led directly to a new ability on the part of so many of us to understand the importance of protecting the earth against the insults of pollution and carelessness that are now posing new threats to the ecological system of the entire earth."

— Vice President Al Gore

Part II

THE PATH

Chapter 2

THE PATH

The environmental problems of greatest immediate concern have changed over the past quarter of a century, and the technologies required to address those problems have changed as well. In the 1970s, environmental protection focused on "end-of-pipe" equipment for controlling air and water pollution. In the 1980s, the physical cleanup of waste sites received particular attention. Today, environmental protection is beginning to involve changes in the fundamental ways our energy, food, fiber, shelter, and consumer goods are produced. The emphasis has shifted from the control and remediation of pollution to the avoidance and monitoring of many kinds of environmental harm. A brief look back at where we started and where we are today can help us define where we would like to be in the future.

Earth Day 1970

5:15 A.M., April 22, 1970: A small group of people gathered on the summit of Cadillac Mountain, the highest point on Maine's eastern shore. They inaugurated a celebration that would sweep across the nation during the next 24 hours and draw more than 20 million people. In elementary and secondary schools from coast to coast, 10 million schoolchildren took part in "teach-in" programs. In Omaha, Nebraska, high school students collected and stacked 156,000 beer and soft drink cans. In the nation's capital, Girl Scouts in canoes picked up garbage from the polluted Potomac River. In New York City, more than 100,000 office workers poured onto Fifth Avenue, and Wall Street was blocked by a rally organized by students from New York University's business school.

Many corporations joined in the celebration. Continental Oil Company announced it was introducing a cleaner gasoline. Scott Paper vowed to invest $36 million to clean up pollution at a pulp and paper mill. And Reynolds Aluminum dispatched trucks to colleges and universities in 14 states to collect aluminum cans for recycling at half a cent per can.

> ***"The environmental problems of the future will increasingly spring from the wonders of 20th-century technology. In the future, technology assessment must be used to understand the direct and secondary impacts of technological innovation."***
>
> *— Council on Environmental Quality, First Annual Report, 1970*

The same day, the State of New Jersey created an environmental protection agency and the Maryland legislature passed 21 bills and resolutions enacting environmental controls. Also on Earth Day, the National Oceanic and Atmospheric Administration was established with the responsibility of integrating a variety of ocean and atmospheric research missions under one authority. Two months later, the President transmitted a message to Congress that combined the functions of five federal agencies into a new U.S. Environmental Protection Agency.

In 1970 we were a nation of 203 million people. Our environmental problems were severe and wide ranging. Eight years had passed since the publication of Rachel Carson's *Silent Spring*, and concern over pesticides such as DDT remained persistent and widespread. Likewise, air pollution was being implicated as a chief factor in the doubling of deaths from emphysema and chronic bronchitis that had occurred between 1956 and 1966. The Santa Barbara oil spill in 1969 had heightened the public's awareness of the sensitivity of marine ecosystems and the dangers of offshore oil drilling. Finally, some scientists warned that the release of toxic substances and other environmental problems would worsen as the nation's economy grew and its population expanded.

As environmental awareness grew, there was a corresponding rise in environmental activism. In 1970 few people belonged to any form of organized environmental movement. Within 10 years, approximately 5,672,000 had joined the 17 largest national environmental organizations. Membership in the Sierra Club had doubled in the two years that preceded the 1970 celebration.

Out of the flurry of activities following the first Earth Day crystallized the outlines of a fundamentally new notion of the relation between technology and the environment. Up to that point, most Americans had little concern about the environmental consequences of the technologies used to produce our goods, energy, and food. The early 1970s saw an increased awareness of the need to anticipate technology's impacts.

The idea that technology could create a host of environmental problems helped give rise to the Office of Technology Assessment, a support agency to the U.S. Congress. It was also a factor leading to the extensive use of environmental impact assessments, as prescribed in the National Environmental Policy Act of 1969.

Another view of technology was emerging, however, that was to have significant business implications. Technology's potential as protector and restorer of the environment was creating a major new sector of the economy. On January 19, 1970, the *Wall Street Journal* reported that the cost to clean up America's environment could result in a $300 billion industry by the end of the century. Three months later, *Forbes* magazine predicted that ". . . the effort to clean up the environment is going to be profitable to those who provide ways and means, systems and cement and pipe and chemicals and just about every other thing you can think of." Pollution-control stocks were hailed as the newest fad on Wall Street. In many ways, environmental technology was born in the first part of the 1970s. By then, environmental technologies accounted for approximately $10 billion in revenues.

At that time, regulation appeared as the primary means of achieving environmental goals. A 1970 survey of the chief executive officers of top Fortune 500 companies indicated that 57 percent wanted to see the government step up its regulatory activities to ensure a level playing field. The growing number of regulations played a key role in creating the demand for environmental technologies. Entire industries developed because of the regulations to control pollution. One example was the manufacture of scrubbers to remove pollutants from industrial and utility smokestacks.

The policy decisions in the 1970s that set the stage for our nation's regulatory framework spurred the development of environmental technologies. But they also had unforeseen impacts that created a number of barriers to the growth of the environmental technology industry. Policy decisions during the present decade will have equally far-reaching impacts over the next 25 years. The Americans who celebrate Earth Day 2020 will look back to Earth Day 1995 as a new beginning, provided we do our job well.

Earth Day 1995

Over the past 25 years, the quality of the national environment has improved because of activities, ideas, and commitments that originated on the first Earth Day. During that time, environmental technologies played an important role in reducing the impact of the U.S. population as we added more than 60 million people to our country.

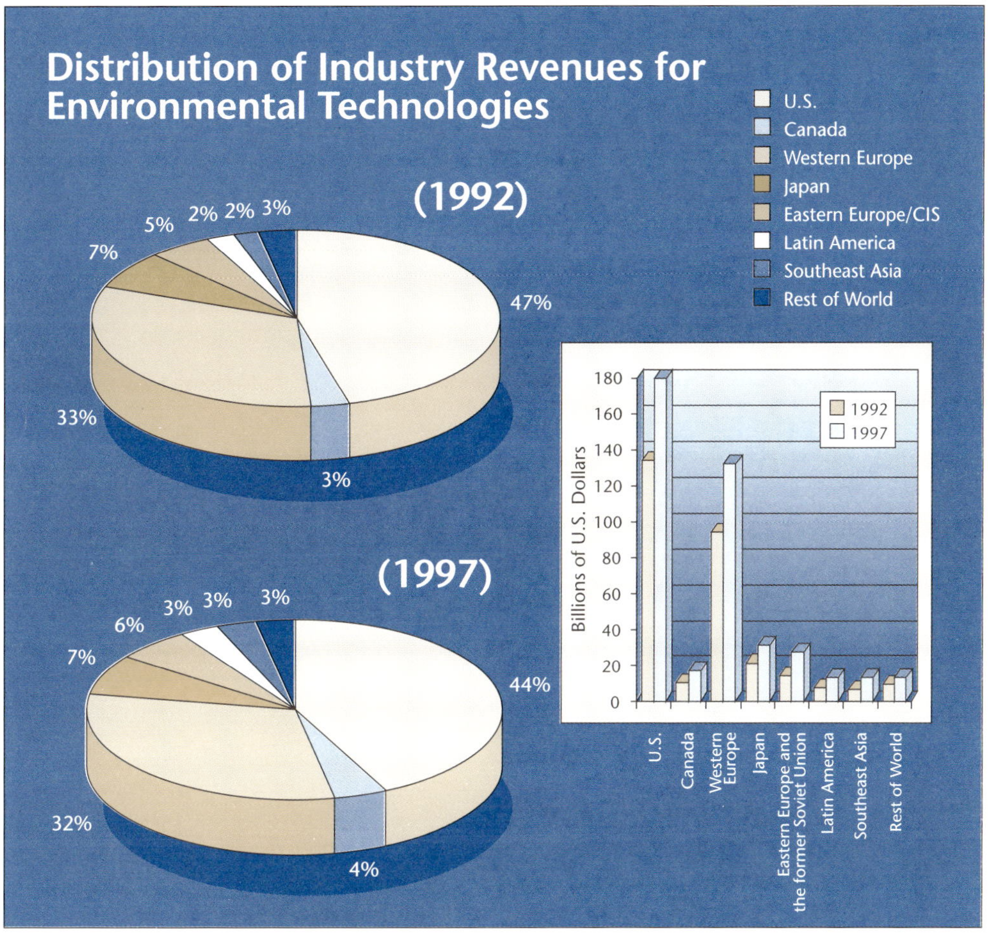

The U.S. market is, and will continue to be, the largest single market for environmental technologies. However, by 1997, growth in certain regions and nations such as Latin America, Canada, and Eastern Europe and the former Soviet Union will outpace growth in the United States.

"Environmental protection sometimes is portrayed as conflicting with economic growth. In recent years, however, governors, state economic development and environmental protection officials, and the business community have recognized that the commercialization of environmental technology offers tremendous opportunities for business creation and expansion and creates high-quality jobs."

— Governor Bob Miller
Former Chairman
Committee on Natural Resources
National Governors' Association

Our manufacturing output is now twice as high as it was in 1970, and many products that did not even exist 25 years ago are being manufactured in mass quantities. Today, eight out of ten Americans work in the service sector, which now accounts for 75 percent of our country's gross domestic product. Our agricultural output has increased by 75 percent since 1970 with no increase in the number of acres under cultivation. Explosive growth occurred in our information infrastructure throughout the 1980s, allowing massive amounts of information to travel between people, among organizations, and across borders. The importance of trade to the U.S. economy greatly increased in the 1970s.

From the U.S. environmental industry's modest start in the early 1970s, a $147 billion business has emerged that employs more than one million people. The domestic environmental technology industry is now larger than many other industries considered key to our economic growth, such as computers, plastics, and pharmaceuticals. The global market for environmental goods and services is approximately $300 billion and is expected to grow to $400 to 500 billion by the beginning of the next century. If one includes recent estimates for investments in energy infrastructure in developing countries, this figure grows to more than $1 trillion by the end of the decade.

Presently, the environmental industry is dominated by service providers, which account for 74 percent of the total revenues. Most of these services are focused on the cleanup of past environmental problems. Though the industry as a whole has grown slowly over the past three years, the sector dealing with prevention technologies increased by almost 15 percent in 1992 and 1993. Other high-growth areas include technologies for resource recovery and energy conservation.

Federal investments in environmental technologies have increased from approximately $550 million in 1970 to $4 billion in fiscal year 1994. Though the macroeconomic impacts of such investments are difficult to estimate, a recent study by the National Commission for Employment Policy shows that environmental technologies that increase resource-use and energy efficiencies can make many firms more productive and competitive, allowing them to grow and create jobs.

Looking back over the 25 years since the first Earth Day, we have reason for optimism and cautious reassessment. Investments in environmental technologies have demonstrated their success in dealing with many environmental problems, but others have proved more immune to technological interventions to date. Since 1972, a national investment of $350 billion for control of water pollution has helped restore water quality in many places. Today, 56 percent of our nation's rivers, 43 percent of our lakes, and 56 percent of our estuaries can fully support fishing and swimming. Since the first Earth Day, we have eliminated many releases of lead into our

A Changing America, 1970 to Present

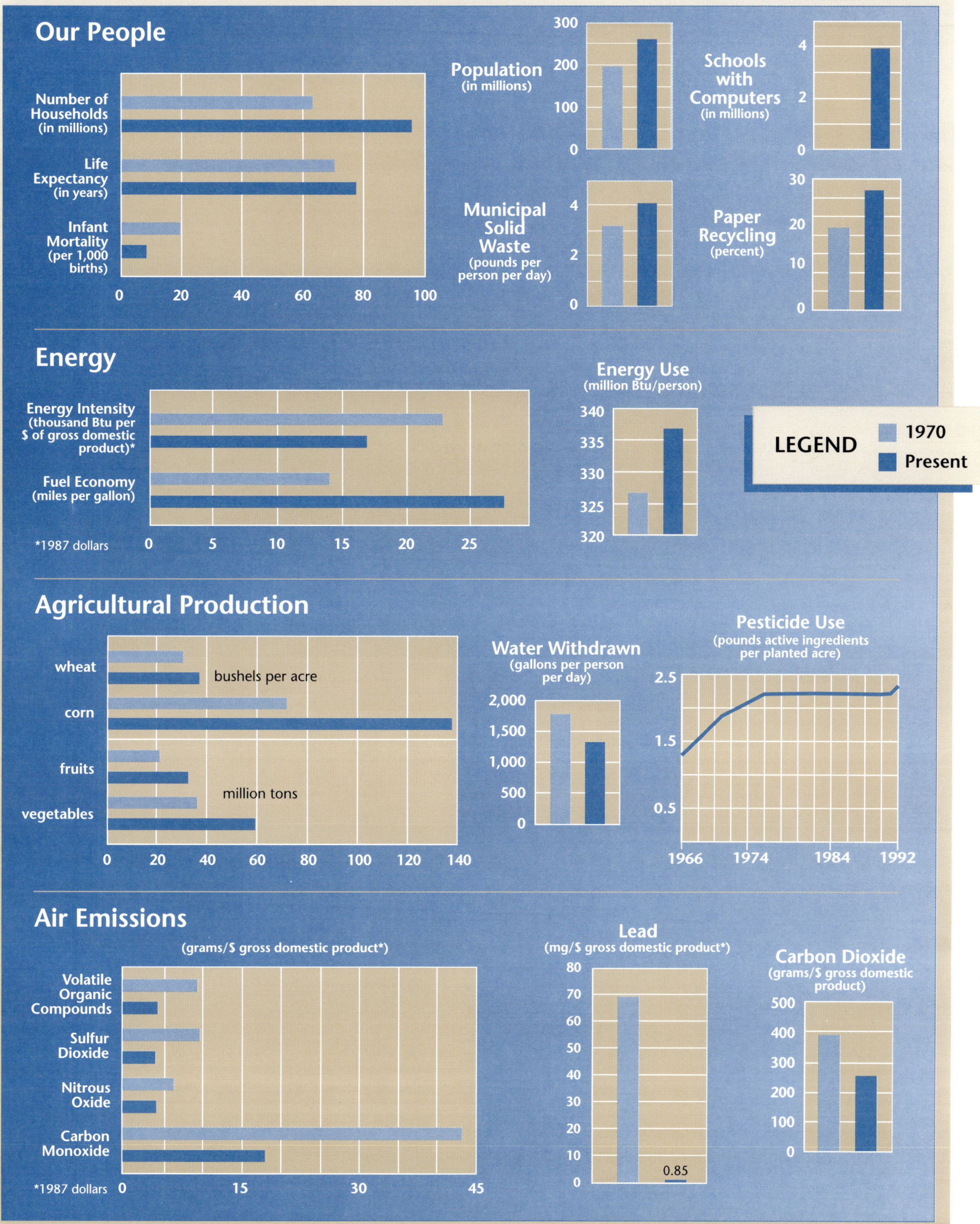

environment and have significantly reduced emissions of sulfur dioxide and carbon monoxide.

In other areas, advances in technology have barely offset our nation's expanding population and changing consumption patterns. The United States remains the undisputed world leader in the production of solid wastes. Each year Americans dispose of 247 million tires, 2 billion razor blades, and 1.6 billion pens. Per capita generation of solid waste has increased by 65 percent in the 25 years since the first Earth Day. Even if we continue to reduce solid waste and aggressively recycle, it is projected that we will merely maintain our position, remaining at a rate of 4.3 to 4.4 pounds per person per day by the year 2000.

As the 1980s came to a close, a new generation of environmental problems — not amenable to end-of-pipe controls or improved disposal techniques — challenged scientists and policymakers. Problems such as global warming, ozone depletion, loss of habitat, and biological diversity are global or regional in scope and result from the use of materials, energy, and land by society at large, not just emissions from the pipes and stacks of industrial facilities.

Unfortunately, many of the gains in environmental quality that we take for

Over the past 25 years we have moved from end-of-pipe technologies used to control pollution to more integrated and systemic approaches such as whole-facility planning and industrial ecology designed to prevent environmental harm. This evolution will continue as we seek additional gains in efficiencies by linking and integrating technologies across processes, industries, and sectors of the economy.

granted are not shared in many other countries of the world. Six million people in India die each year from acute respiratory infections caused by smoke from small cooking fires and heating appliances. Traffic jams in Bangkok cost $1 billion a year in medical expenses and reduced productivity. In addition, 35,000 children under age five die worldwide every day as the result of preventable diseases. Most of these deaths are associated with unsanitary living conditions. Finding appropriate solutions for these problems will remain one of the great challenges of the coming decades.

"Committed corporate environmental stewardship, linking operational efficiency with innovative daily procedures, is good business in the 1990s."

— Tedd R. Saunders
Saunders Hotel Group
Eco-Logical Solutions, Inc.

Looking back over the last 25 years, we have a vantage point from which we can explore the evolution of environmental technologies. We can begin to see the outlines of a transformation in environmental quality, technologies, management practices, and policy. Where did this transformation begin, where is it heading, and where will it take us?

Given the severity of environmental problems faced in the 1970s, the easiest and fastest technological solution was to control emissions with add-on devices. This first generation of technologies for pollution control proved very successful in initial reductions of releases of pollutants into water, air, and soils. They had little or no impact on transforming existing systems of production, which were characterized by high energy and materials requirements and relatively low efficiencies.

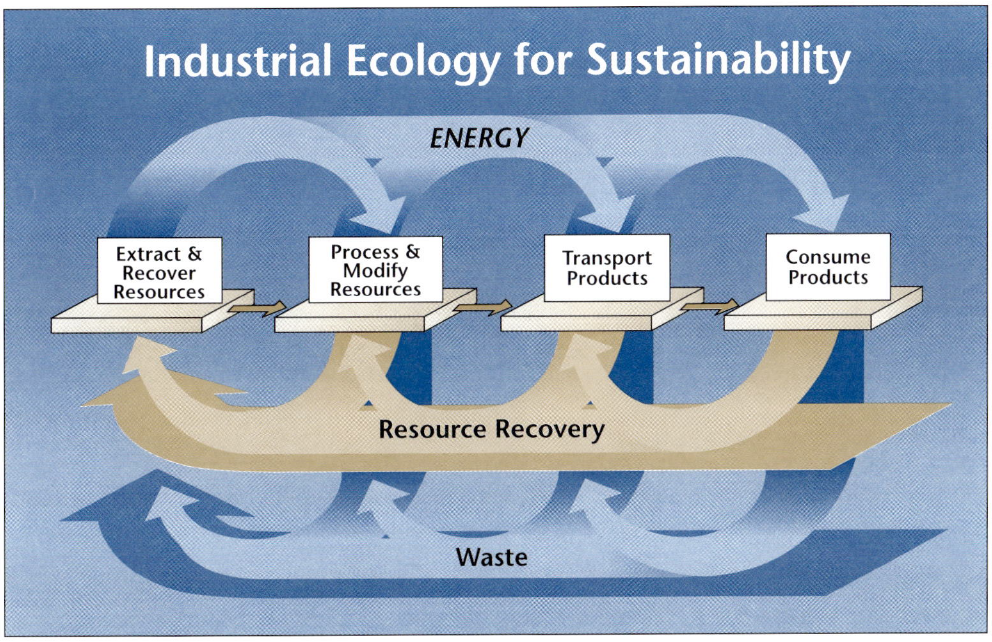

Using the principles of industrial ecology, we seek to design systems for producing energy, food, or products that minimize the use of valuable resource inputs and reduce waste outputs. Industrial ecosystems are built on an understanding of how energy, materials, and chemicals flow through these processes. Opportunities are sought to minimize energy and material use, re-use wastes, and facilitate the efficient re-use and recycling of end products.

A number of companies, however, were beginning to think about manufacturing processes, their interrelationships, and their aggregate impacts on the environment. Process integration approaches proved extremely successful at further reducing environmental emissions while simultaneously improving energy efficiency and product value (see figure on page 14). For instance, through process changes, Intel, at its Aloha, Oregon, plant, was able to increase its production 2.5 times with no increase in emissions without investing in traditional control technologies. The key was integrating environmental considerations into the design of new processes.

In the early 1980s, many firms took another step down the path toward improved environmental management. They began to redesign their whole facilities

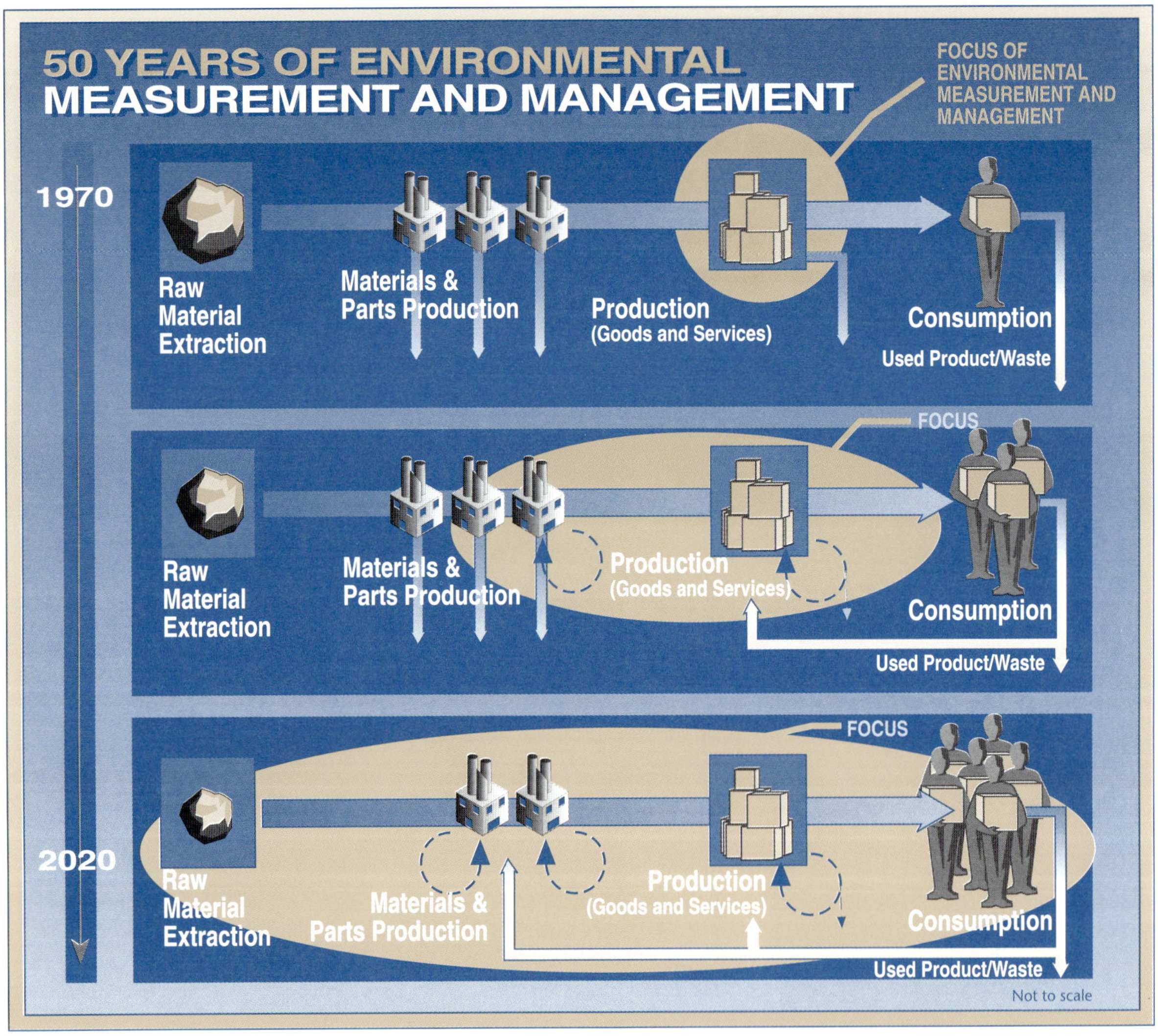

Increasingly, businesses of all types are looking beyond the borders of their facilities for new opportunities to increase product quality, reduce costs, and lower environmental impacts. This often means working directly with suppliers of raw materials or product parts and responding to the needs of customers and clients with more comprehensive product information. The boundaries of environmental management and responsibility will continue to expand, as producers of our consumer goods, services, energy and food increasingly look across the entire life cycle from extraction to production and consumption.

for environmental and energy performance. A number of states tried to encourage this trend and, by 1995, 22 states had comprehensive whole-facility planning laws on the books. Five years ago, Hyde Tools in Massachusetts decided to get ahead of the curve environmentally by using a whole-facility planning approach that integrated environmental objectives into mainstream corporate decisionmaking. In the words of Doug DeVeries, who heads up Hyde's environmental program, "If we get ahead of the regulations, we do not have to worry about them." Over a five-year period, Hyde Tools spent $350,000 in new machines and process improvements to meet or exceed environmental standards. These process changes saved Hyde more than $900,000 — almost a three-dollar return for each dollar invested.

In the mid-1980s, a new paradigm emerged constructed on the premise that industrial systems can be modeled after ecological principles. "Industrial ecology" emphasizes the need to understand the flow of materials and energy through industrial systems, the effects of these flows on the environment, and the influences of technology, regulations, and operational practices on the flows. Industrial ecology incorporates such approaches as analyzing process flows, recycling wastes into inputs, examining the entire life cycle of products, and designing products from the beginning with the environment in mind.

Underlying this technological transformation is a subtle, but important, shift in the way firms view their environmental responsibility, measure their impacts, and interact with their suppliers, customers, and neighbors. Of large firms responding to a recent Price Waterhouse survey, three-fourths indicated that they have environmental auditing functions in place, representing an increase of 40 percent from 1990. More than 11,000 U.S. firms now submit annual environmental information to the Securities and Exchange Commission.

Expanding the Circle of Environmental Management

More and more businesses are working with suppliers and customers to increase opportunities for environmental management. Home Depot, a $12.4 billion home center retailer with 340 stores in the United States and Canada, leverages its buying power to seek products that minimize the burden on the environment. Moving upstream, the company uses an independent scientific service to evaluate manufacturers' claims such as "biodegradable," "recyclable," and "saves energy" before buying such products for their stores. Home Depot also passes this information on to consumers , enabling them to make more environmentally informed choices. Another example is the 3M Company and Akzo Nobel, Inc., which recently developed an environmental guide to help one large user of their products, the furniture industry, increase environmental awareness, and minimize compliance costs.

Energy: A Critical Technological Challenge

Our homes, appliances, lighting, heating, air conditioning, and communications systems are all powered by energy in various forms. Electricity is the second largest energy expenditure of the average home. The fraction of household income spent on electricity remains at about 2 percent, and the Department of Energy estimates that level will be maintained in the future. To produce energy economically, the U.S. power industry expanded its generation capability over the past 25 years using indigenous coal resources and nuclear energy. Demand for residential electricity and the generation of electricity from coal have increased significantly over the same time period and are projected to continue to grow as the number of households, businesses, and industries increases.

Producing electricity from coal has an impact on our environment. Combustion of fossil fuels releases into the atmosphere small amounts of sulfur dioxide (SO_2), nitrogen oxides (NO_x), carbon monoxide (CO), and particulate matter (PM) along with large amounts of carbon dioxide (CO_2). Since 1970, many end-of-pipe technologies successfully decreased the emissions released per unit of power delivered. This allowed many utilities to comply with the Clean Air Act while still providing affordable electricity. Although gains have been made in controlling these point source emissions, overall emissions have not declined because of increasing energy consumption. Further gains in emissions require continued improvements in the efficiency of energy production and use, as well as a shift to production technologies that produce few or no emissions. Technologies and processes will need to be developed to control or convert CO_2 emissions that are not addressed by current technologies.

U.S. Residential Electricity Use

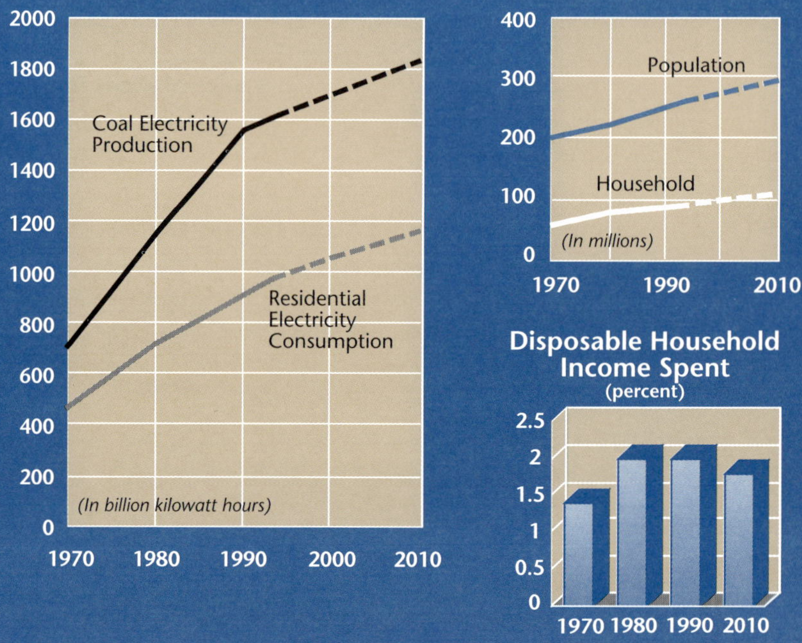

Emissions from Coal-Generated Electricity
(pounds/thousand kilowatt hours)

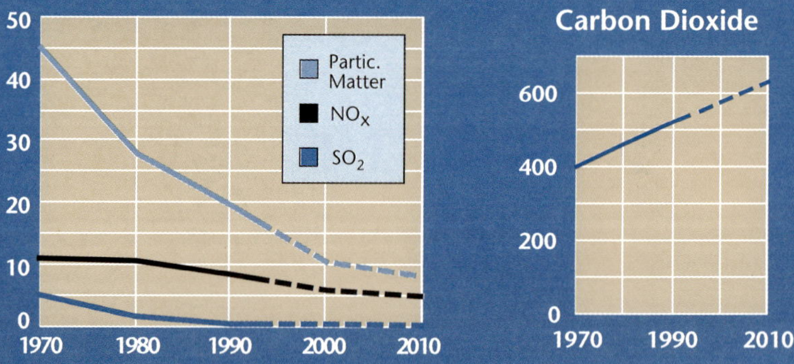

Total Air Emissions
(million tons)

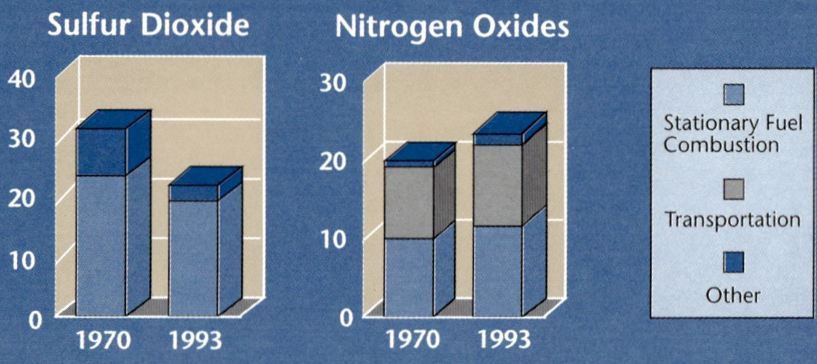

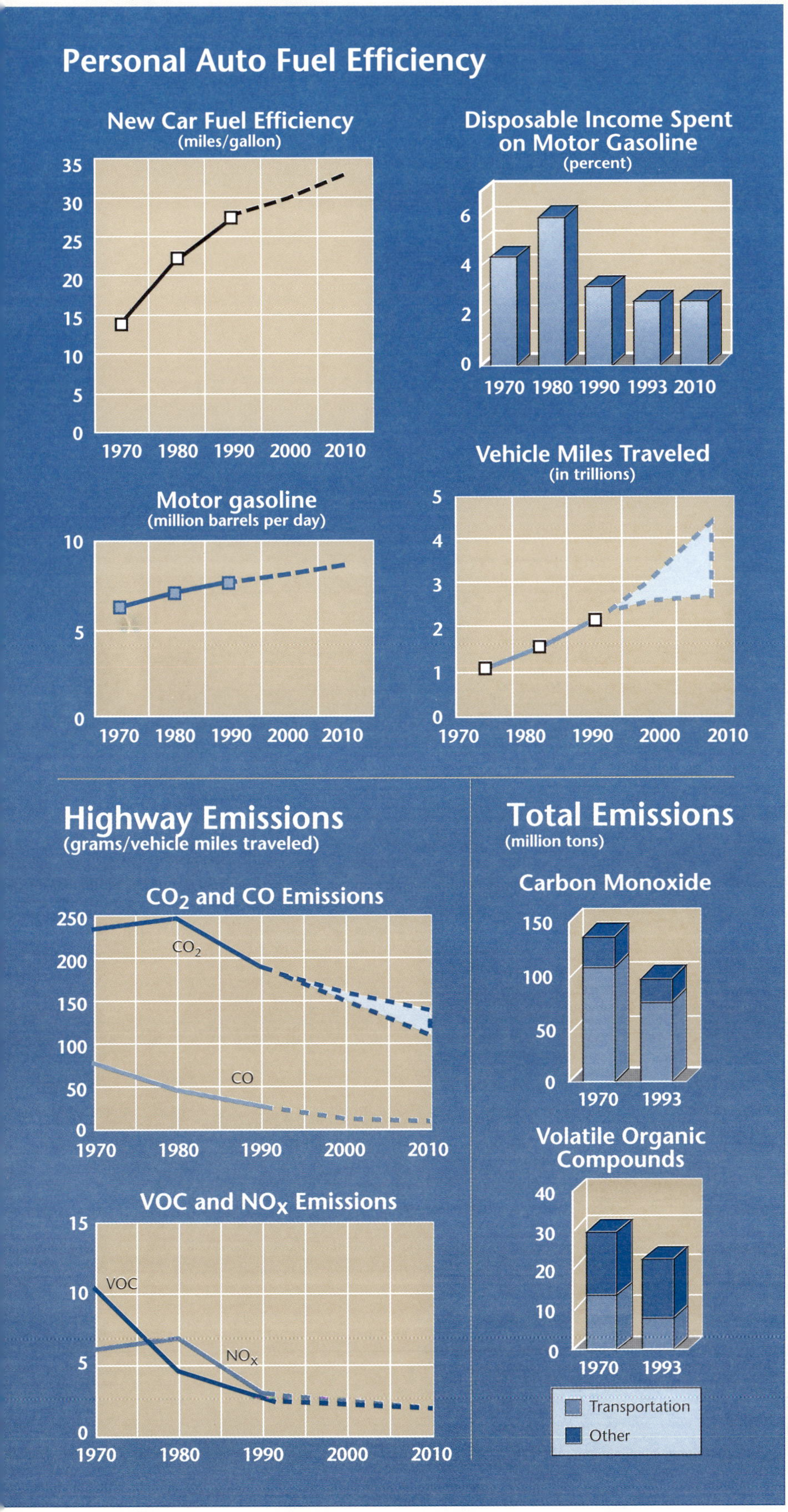

For most of us, a major share of our household energy budget is directed to automobile fuel. Although the fuel economy of new cars nearly doubled in the past 25 years, we travel more than twice as many miles today as compared with 1970. More than half of our petroleum is imported, contributing significantly to our nation's trade deficit and resulting in decreased national security.

With the exception of carbon dioxide, technology has significantly reduced the emissions that are produced per mile traveled. NO_x emissions have been more difficult to control using existing technologies. Although overall emissions decreased, progress has been slow because of increases in miles traveled and total gasoline consumed.

There is a clear need for a new generation of vehicles capable of using far less fuel with significantly fewer emissions. The Administration's "Clean Car Initiative," undertaken in partnership with major U.S. automakers, seeks to develop a new vehicle with three times the fuel efficiency of existing cars and no sacrifice in cost, comfort, or safety.

We have witnessed a merger of the total quality approach to manufacturing with corporate environmental management functions. The corporate model itself is changing and the boundaries between firms, between customers and suppliers, and between competitors are becoming blurred. More and more industries and service firms are integrating their upstream providers and downstream customers into their environmental education, measurement, and management strategies.

In this way, they are expanding the circles of product and environmental responsibility and providing new opportunities for waste reduction, energy efficiency, and material reuse and recycling. This transformation will impact manufacturing, food and energy production, construction, and virtually all other sectors of our economy over the coming decades.

Earth Day 2020

The shifts taking place in the 1990s in our approaches to measurement, production, and management will have profound and far-reaching impacts on the development and use of environmental technologies. Our vision of the future is built on understanding and accelerating these environmentally beneficial trends. Implementing changes in complex technological systems often requires years, or decades, so we need to act in anticipation of where we would like to be. The path of

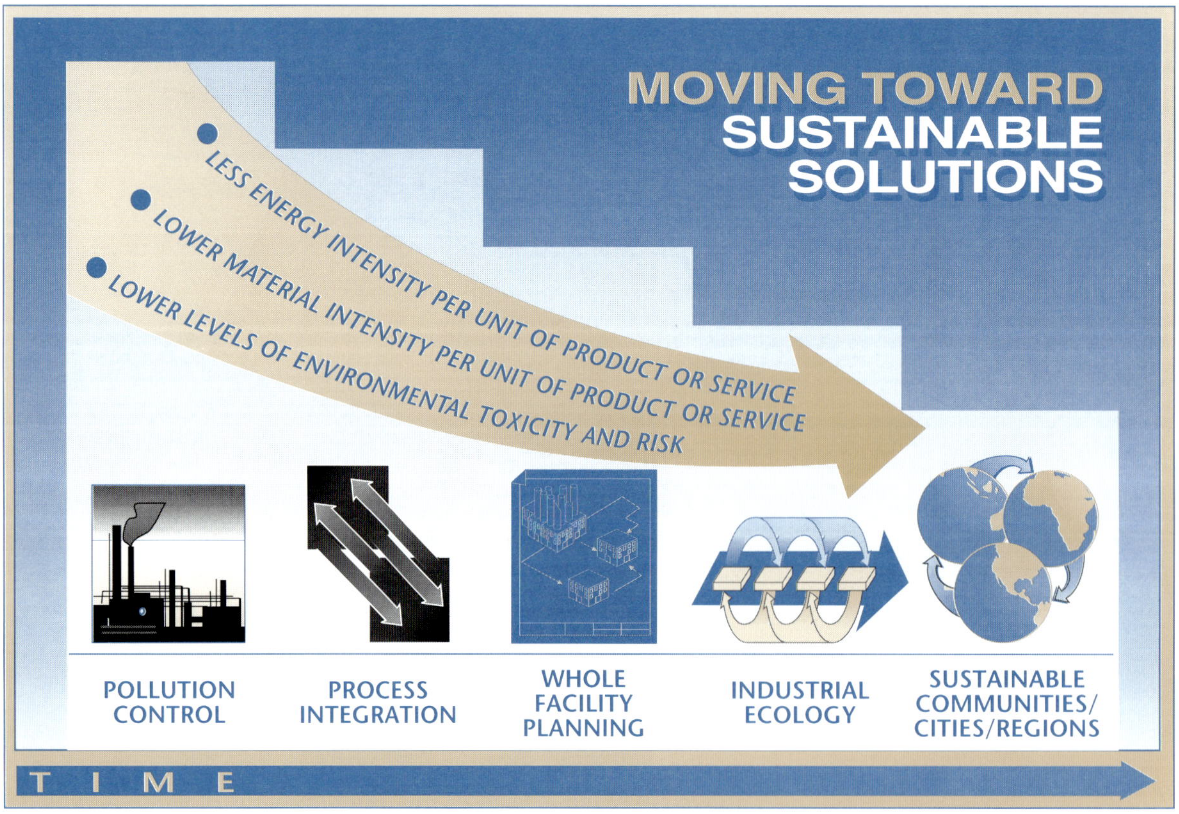

The shift to more sustainable technologies will mean significantly reducing the amount of energy and materials we use in producing our goods and services while decreasing risks to humans and the environment.

technological development is being set by today's policies, investments, and decisions. We cannot predict what surprises or technological breakthroughs await us over the next 25 years, but we can build on our present understanding, goals, and aspirations to help shape our common future.

The policy and technological implications of this transformation will be profound. We will move from discrete technologies tackling particular problems to the integrated design of large systems capable of reaching much higher levels of efficiency. In the end, environmental technologies will virtually disappear as distinct pieces of hardware, and environmental concerns will become more fully integrated in the fundamental design of all products and processes.

These new systems will mimic the operation of natural systems, where "wastes" are not only minimized, but also are used and reused as resources. These systems will be built on scientific advances in areas such as computing, genetics, materials, biotechnology, miniaturization, and modeling. Over time, the principles of industrial ecology will be extended beyond the industrial sector itself to encompass the larger systems by which energy, food, fiber, transportation services, and housing are produced. These principles help facilitate the creation of sustainable communities, cities, and regions.

"Together we must transform our technological infrastructure to use less energy and less materials, to cause less environmental harm, and, at the same time, to provide those goods and services that we all desire. In this joint endeavor, we seek more than small steps down the familiar technological paths. Instead, we seek a transformation of our industrial economy."

— John H. Gibbons
Assistant to the President for Science and Technology

It is this transition of technologies that will allow us to provide more products and services while decreasing environmental risks and reducing requirements for energy and materials. However, given current projections of global population and economic growth, the challenge to "do more with less" using new technologies is considerable. Some researchers argue it will be necessary — and possible — to achieve a tenfold increase in resource productivity in industrialized nations over the next 30 to 50 years. Therefore, any future transformation of technologies must be accompanied by a significant transition to higher levels of efficiency.

Changes of this magnitude will require regulators and the regulated community to move from command-and-control regulatory systems to cooperation and partnerships. The levels of technological innovation needed to meet future environmental challenges can only be achieved if we work together to develop long-term goals, measure performance along multiple dimensions and scales, and implement complementary, mutually reinforcing policies that encourage high levels of innovation. If we can do this, significant change is possible.

By Earth Day 2020, Americans will be living in a more sustainable economy and will regard their skills in environmental protection and management as the key to continued economic growth and international competitiveness. Our goal is to generate 40 to 50 percent less waste, use 30 to 40 percent less energy, and 20 to 25 percent less materials per unit of gross domestic product.

The relationships among companies and between companies and the community at large will be more cooperative and productive. Companies will be co-located to share resources more effectively. Sophisticated environmental auditing systems combined with new integrated efficiency measures will continue to increase productivity while giving companies the strategic insights and flexibility to identify and design innovative technological solutions. We will have developed new ways of organizing production that stress the participation of workers in central decisions affecting the environment, their working conditions, and long-term economic welfare. U.S. industries will be major exporters of clean technologies for manufacturing and worldwide leaders in environmental services.

As urban and industrial systems change, minimization of resource use will be an integral part of the design of every product and manufacturing process. Innovative ways to recycle waste streams will now be commonplace, and secondary markets for the byproducts will boom. Products will be designed with their life cycles in mind. Toxic emissions will be reduced or eliminated, and the use of materials will be closer to achieving closed loops and thus more sustainable. Entirely new products and services will have been developed with product longevity, reusability, and intensity of product use as primary considerations. The prices of products will more accurately reflect their costs to the environment and society.

In 2020, we will have achieved integrated collection of the information that flows from monitors and sensors at global to molecular levels. The information will be widely available for analysis using network-based visualization tools, geographic information systems, and simulation models. The expansion of the national and global information infrastructure will support new forms of collaborative work among individuals, organizations, communities, and nations. Decisions on environmental problems and solutions will be based increasingly on interdisciplinary assessment, international research, and mutual understanding.

Information that is necessary for critical design trade-offs for products and processes will be widely available. Monitoring of manufacturing processes will help companies use resources more efficiently and make necessary adjustments to maintain high productivity and standards.

Communities will excel in the design of quality environments and efficient infrastructures for businesses, residents, and recreation. We will have achieved a sophisticated level of environmental awareness in all sectors of society. In 2020 people will have become an integral part of the environmental management process. Information based on technical and scientific knowledge will be available and

Environmental and Technological Progress in the Computer Industry

Groups such as the Institute of Electrical and Electronic Engineers and the Electronic Industries Association have been supporting the electronics and computer industries in the search for new ways to manufacture tomorrow's computers using fewer materials, less energy, and with less impact on the environment. Circuits are now 1,000 times thinner than a human hair and a single handheld device has the processing power of a mid-range computer of the 1970s. While the industrial revolution enhanced productivity by a factor of about 100, modern microelectronics has already enhanced productivity in information-based technology by a factor of more than a million. The industry is increasingly striving to find the best technical and environmental options for materials, design, and power consumption.

understood by an increasing proportion of the population.

By 2020, we will have achieved high levels of environmental protection and decreased health care costs due to some pollution-related chronic diseases. We will be able to monitor human exposure to pollutants with more precision, using biological markers.

We will have new methods of restoring and enhancing ecological systems that have been degraded over the preceding decades and centuries. Increased ultraviolet radiation and ozone depletion could still be a problem, however, and we may be facing global climate changes.

In 2020, the United States will produce sufficient energy to support an additional 30 to 40 million people with significantly less emissions of harmful substances than in 1995. We will have shifted to energy sources that are more environmentally friendly, developed advanced energy management and storage systems, and increased the efficiency of energy generation technologies. U.S. industry will have significantly increased the efficiency of its energy use. A wider range of renewable energy resources will be in use, including high-efficiency power sources based on fuel cell technologies that transform hydrogen from a variety of sources and oxygen from air into water and energy with zero-emissions. The U.S. clean energy industry will be a major world exporter. Renewable energy resources will increasingly power communities throughout the world. Buildings and homes will have high-efficiency central lighting just as we have central heating and cooling today.

"I believe that this first step of creating a consensus about the long-term vision or strategy for the United States will have a very beneficial impact for the world as a whole."

— Ismail Serageldin
World Bank

More of the average American's food in 2020 will be grown by farms located near consumers. Farmers will belong to a broadly diversified agricultural business community. From 1995 to 2020 they will have shifted from single crop commodity production to specialty crop and integrated farming systems. The electronic age

farmer will have increased the use of information systems to guide crop selections and production decisions. Computerized farm equipment using global positioning systems and remote sensing will facilitate precision farming with a greater reliance on environmentally benign fertilizers and pest control methods. In general, farm equipment will be less polluting, crops will be adapted to a wider range of climactic conditions, and farmers will produce a wider variety of food, fiber, energy, and crops for industrial use while using soils, water, and fertilizers efficiently.

Is this world achievable? If we work together in partnership — the private sector, government at all levels, academia, nongovernmental organizations, and the public — by Earth Day 2020 we will see a more productive society using technology wisely to achieve a sustainable future. The following chapter details some of the steps we can take together to reach this vision.

Part III

MOVING FORWARD

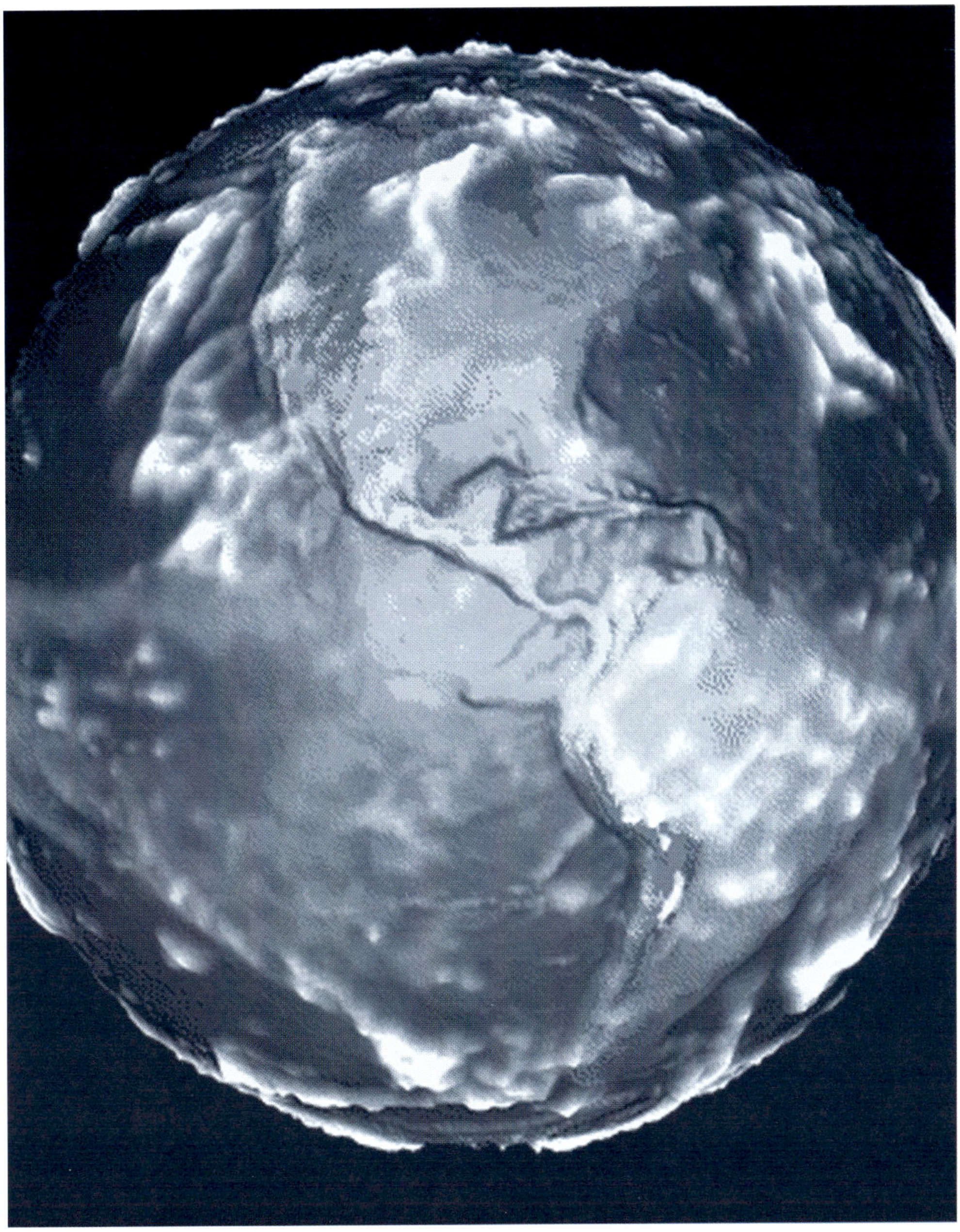

Chapter 3

MOVING FORWARD

The future just outlined is achievable. We can lay the foundation today for a set of policies and programs that will establish a new course for the development and use of environmental technologies into the next century. We must, however, broaden our environmental tool kit, replacing those instruments that are no longer effective with a new set of tools designed to meet today's challenges and tomorrow's needs.

Over the past six months, thousands of people have discussed the merits of setting a new course and identified the tools we will require to change our present direction. From these discussions, a core set of five themes emerged that can be used to guide future activities. These themes are designed to establish the framework for partnerships, goal setting, policy development, and action.

Performance, Flexibility, and Accountability

The vision outlined in the previous chapter provides a challenge to all businesses, large and small, and all sectors of the economy to improve their environmental performance and use less energy and fewer materials in providing products and services. Together, we need to develop a new generation of incentive-based policies and programs that stress performance, flexibility, and accountability. Our focus must shift to systemic whole-facility, multi-media, and multi-sectoral approaches. These programs need to be credible and inclusive, allowing the participation of all interested parties in the application of new technological solutions and the information and organizational structures that support them.

Innovation for Environmental Results

Achieving the vision is possible only if scientists, engineers, and entrepreneurs provide the innovations to move us from an environmental paradigm built on control and cleanup to one also based on efficient resource use and avoidance of environmental harm. Over the long term, research, development, and demonstration programs should facilitate a broadening of focus from *react-cure* to *anticipate-prevent* policies and allow our economy to do "more with less." As our focus expands, however, we must ensure that pressing remediation problems are dealt with and cost-effective compliance with existing regulations is achieved. Government programs need to facilitate and accelerate private sector research, development, and commercialization efforts. Programs to demonstrate new technologies must be market driven, linking solutions developed by the public and private sector with market needs.

Commercialization

The vision cannot be realized if good ideas remain on the shelves or in the back rooms of developers. We need to bring together innovations with the necessary capital and market information — and do it in a timely manner, especially for small businesses. To accomplish this objective, we will need a wide range of complementary policies to support investment and the diffusion of successful technologies. We also need to commit to building environmental awareness and institutional and technological capacity in other parts of the world to foster the use of appropriate environmental technologies.

Sustainable Communities

The end point of our vision must be healthier, environmentally sound, and economically viable communities. These communities, both urban and rural, will need to make significant advances toward sustainability over the coming decades, using environmental technologies appropriately and wisely. New generations of more efficient and integrated infrastructures for energy, transportation, water and wastewater treatment, and communications must be carefully designed today to ensure a more sustainable technological base for tomorrow's cities, towns, and neighborhoods.

Learning and Working Together

Our vision must be built on more effective, open, and productive collaboration among stakeholders. It will require new models of gathering, sharing, and analyzing information; coordinating work; and educating and training professionals, policymakers, and the public at large. The theme of working together underlies the other principles discussed above.

The following sections will expand on these themes. Each section is organized into the following parts: a **finding** that explains the primary issue or issues that emerged during discussions with stakeholder groups; a **goal** that describes an end point or directional change necessary to address the issues presented in the finding; and, in some cases, **initiatives** that outline specific actions that will be taken to achieve the goal. Throughout this chapter we include summaries of observations and recommendations that came directly from the participants in one or more of the workshops or conferences held around the country.

Performance, Flexibility, and Accountability

Incorporating greater incentives for innovation and avoidance of environmental harm into federal and state environmental policy is one of the central components of this national environmental technology strategy. The policy and regulatory framework administered by the federal government and state agencies creates much of the commercial demand for environmental technologies. Traditional command-and-control approaches, however, provide few incentives for technological innovation or for taking actions beyond those required to meet a particular standard. Performance standards in concert with market-based incentives encourage innovation and the development of cost-effective compliance strategies.

FINDING **Although federal and state environmental regulations define and shape the market for environmental technologies, their inflexibility and unpredictability in some cases limit the development, commercialization, and application of new environmental solutions.**

GOAL **Achieve continuous improvement of the environmental performance of U.S. industries, using the most advanced technologies and cost-effective means possible, by strengthening incentives for innovation within the regulatory system.**

The federal government will work closely with the states to stimulate the development and adoption of environmental technologies by strengthening the incentives for innovation within regulatory, permitting, compliance, and enforcement programs, and by identifying and reducing barriers to innovation in these programs. EPA, working with the states, is developing a new oversight policy for delegated state environmental programs that rely on joint strategic planning, collection and publication of a core set of environmental performance measures, and identification of shared environmental problems. Implementation will be through a joint EPA/state agreement that matches the level of oversight with the unique problems and strengths of the state program.

The President and Vice President recently announced a series of policy changes and initiatives at the federal level to reinvent environmental regulation. In calling for a "government that offers opportunity, demands responsibility, and shrinks bureaucracy," the President outlined a series of high-priority actions that will improve the current system and generate the building blocks for the regulatory system of the future. Key improvements of the existing system focus on performance- and market-based regulation; setting priorities based on sound science; building partnerships; cutting red tape; improving access to environmental information; and promoting better accountability, compliance, and enforcement. Alternative performance-based environmental management strategies, third-party auditing for

STAKEHOLDER VIEWS

The current federal and state regulatory framework lacks incentives for innovation and pollution prevention and includes many disincentives to the development and adoption of new environmental technologies. Federal and state regulations are often unpredictable, complex, duplicative, uncoordinated, overly prescriptive, and inflexible. Government needs to set flexible, performance-based requirements and "get out of the way." However, there should be no flexibility without accountability. As a nation, we must build a new consensus around results rather than methods.

Promoting Innovation through Voluntary Actions by Industry

Increased reliance on self-designed, voluntary initiatives by companies can achieve significant reductions in pollution levels and often in manufacturing costs as well. Several voluntary pollution prevention programs are administered by the U.S. Environmental Protection Agency with the support and cooperation of other federal agencies. Each offers technical guidance, referrals to sources of assistance, and recognition for participants.

An example is the 33/50 Program, which encourages industry to reduce emissions voluntarily by 33 percent in 1992 and 50 percent by 1995 from 1988 levels. Another is the Climate Wise program, which helps companies find innovative ways to reduce emissions of greenhouse gases. The Design for the Environment program encourages businesses to incorporate environmental considerations into the design of products and manufacturing processes.

The Energy Star Buildings program helps owners of commercial and industrial buildings cut energy costs. The Energy Star Computers program encourages computer manufacturers to produce energy-saving equipment that reduces power automatically when not in use. The Green Lights program provides on-site technical guidance to encourage companies to switch to energy-efficient lighting. And the WasteWise program promotes cost-effective steps to reduce the amount of solid waste generated by businesses.

industry compliance, and multi-media ("one-stop") permits will be key components of the new regulatory system.

The Initiatives

As part of this environmental technology strategy, the federal government has launched four initiatives to begin the shift from command-and-control policies to a more flexible, less prescriptive approach. Under these initiatives, the states, communities, and individual businesses and industries will take on greater responsibility for finding innovative ways to accomplish environmental objectives.

INITIATIVE **Through Project XL (Excellence and Leadership), the federal government working in cooperation with the states and selected firms will test a new approach to environmental management that will provide incentives for industry to find innovative ways to achieve environmental goals.**

The goal of this initiative is to improve the environmental performance of U.S. industries by providing the option of following an alternative path to existing regulatory-driven compliance. The pilot program will require reductions in discharges below current regulatory standards in exchange for greater flexibility in achieving environmental objectives. This initiative will not involve modifying existing regulatory goals; it will simply provide an alternative that responsible industries and companies could choose on a voluntary basis.

> *"I want a government that is limited but effective, that is lean but not mean, that does what it should do better and simply stops doing things that it shouldn't be doing in the first place, that protects consumers and workers, the environment, without burdening business, choking innovation or wasting the money of the American taxpayers."*
>
> *— President Bill Clinton*

The key elements of this initiative are performance, flexibility, accountability, and partnerships. It will also require long-term goal setting and regular reporting of results. This more flexible approach will offer a win-win opportunity for industry, federal and state regulatory agencies, and society as a whole. The result will be a cleaner environment through a more efficient, less-costly regulatory system. The key components of Project XL are outlined in the box on the following page. Project XL will build on the work and principles of EPA's Common Sense Initiative, seeking "cleaner, cheaper, smarter" environmental protection by working with industry and other key partners.

INITIATIVE **The federal government will work with industry, the states, and environmental groups to evaluate the effectiveness of independent third-party auditing for monitoring the environmental performance of U.S. industry.**

Key Components of Project XL

Review: An extensive environmental review of the facility will be undertaken to establish baseline conditions.

Strategic Planning: The facility will develop an alternative strategy to achieve goals for emission reductions beyond the standards provided by current regulations.

Community Involvement: Community involvement will be a key element of the strategic planning and goal-setting process.

Approval and Implementation: Upon EPA approval, the facility will implement the alternative strategy, which will replace many existing regulatory requirements.

Monitoring and Reporting: The alternative strategy will contain a monitoring and reporting plan to allow auditing of environmental performance at the facility with regular public reporting of progress.

Many companies conduct periodic audits of their operations to determine whether they are in compliance with federal and state environmental requirements. Although these audits are typically conducted by the companies themselves, some are undertaken by consulting firms or other independent environmental experts. Third-party audits, if conducted properly, can provide independent verification of a company's environmental performance and compliance without direct expenditure of government resources. Institution of independent auditing will allow EPA to focus its limited resources on noncomplying facilities. Over the next year, EPA will test third-party auditing with 12 partners from the public and private sectors. Following the test phase, the agency will work with the states, industry, and environmental organizations to determine whether a shift toward third-party auditing on a broader scale is viable. Exchange of information and experiences will be sought with the International Standards Organization and the European Union's program to develop environmental management and auditing systems.

INITIATIVE **The Environmental Technology Initiative will strategically apply resources to facilitate the development, commercialization, and use of environmental technologies. These resources will be focused on reducing the regulatory and market barriers that limit the ability of technology developers to introduce technologies to the marketplace.**

President Clinton in his State of the Union speech on February 17, 1993, outlined a new technology initiative to advance environmental protection through the development of innovative technologies while increasing the global marketability of U.S. environmental technology. To achieve these goals, the President called upon the Environmental Protection Agency to lead a multi-year, interagency Environmental

Technology Initiative. In 1994, 73 projects were funded. The intention of the initiative is to cultivate a regulatory climate that is conducive to innovation. The U.S. framework for environmental management must be adapted to ensure that incentives for the development and use of innovative technologies are strengthened.

INITIATIVE **The Environmental Protection Agency will continue to work with state, tribal, and local governments to streamline existing environmental permitting and approval programs and to devise new approaches that will encourage pollution prevention and technological innovation.**

Traditionally, environmental permits are used to regulate facilities that discharge, treat, store, or transport pollutants. A small company may require dozens of permits—one for each pollutant and each media (air, water, land). Current permit programs are often cumbersome and slow. Generally permitting agencies do not encourage a comprehensive, multi-media approach or the use of innovative alternatives, including new treatment technology and pollution prevention. They also provide little opportunity for the development, testing, and demonstration of innovative solutions. The EPA is working closely with the states to streamline the permitting process and incorporate incentives for pollution prevention through the application of environmental technologies.

In addition, EPA will streamline and enhance its approval programs for monitoring methods and explore performance-based approaches for establishing monitoring requirements. The agency is simultaneously accelerating media-specific processes and developing an integrated approval process.

Innovative Technologies for Waste Remediation

Molten Metal Technology, Inc. has successfully teamed entrepreneurial spirit with technical creativity to provide an innovative, environmentally sound alternative to traditional waste treatment methods for source reduction and recycling. Under this approach, a molten metal bath acts as a catalyst and solvent to convert wastes into their elements. The elements are synthesized into new products of three kinds: gases, ceramics, and metal alloys. This approach completely destroys hazardous compounds, exceeding regulatory standards for emissions and residuals.

A related technology uses a molten metal bath to separate radioactive from non-radioactive elements, reducing the volume of radioactive waste by as much as 1,000 to 1. The radioactive elements are then sealed into a stable form for final disposal, and the remaining substances are recycled.

Innovation for Environmental Results

The potential to improve the environment is directly related to our ability to develop and apply technological solutions. Technological innovation will allow us to respond effectively to environmental problems resulting from past practices and develop and commercialize the new technologies required to minimize or prevent environmental problems of the future. Advances in information and monitoring technologies will also allow us to evaluate our progress in addressing these problems and devise new solutions.

As in many areas of science and technology, the United States excels at the fundamental research that supports the development and commercialization of new environmental technologies. But we often fail to bring these innovations to market. Sometimes innovations are commercialized in other countries, which may result in the loss of potential American jobs. Sometimes innovative technologies are never commercialized at all, which can result in environmental damage or excessive cleanup costs. Under this national environmental technology strategy, the partners will work together to speed the flow of innovations from the laboratory to the marketplace.

RESEARCH AND DEVELOPMENT

FINDING **The federal government plays an important role in funding basic and applied research and development (R&D) that is key to the development of future generations of environmental technologies. The federal government also facilitates private sector and cooperative investments in needed R&D, by reducing uncertainties caused by the regulatory, verification, and permitting systems.**

STAKEHOLDER VIEWS

The federal government should take the lead in funding long-term, expensive, and high-risk research and pilot projects. The federal R&D strategy must balance the urgent need for near-term solutions to remediation problems with long-term research for sustainability. R&D should shift in focus from environmental remediation and pollution control to avoidance and industrial ecology.

The federal government should also develop more effective ways for the public and the private sector, in partnership, to make long-term investments and share development risks. In particular, government should help private companies make use of federal technical expertise and testing sites, verify the performance of environmental technologies, and develop better tools for decisionmaking.

Research and development is a key strength of U.S. society and a major source of the impressive growth of the U.S. and world economies since the end of the Second World War. Although the combined industry and government spending in the United States is larger than that of any other country, total non-defense R&D in the United States is substantially smaller relative to the size of our economy than it is in Germany or Japan. In the United States, it is 1.9 percent of gross domestic product, compared with 2.7 percent and 3 percent in Germany and Japan, respectively. In addition, federal R&D investment has slowed over the past 10 years. At the same time, more and more industries have focused their research laboratories on near-commercial technologies or eliminated their R&D investment altogether. This approach promotes the transfer of technology into new products and enhances short-term competitiveness, but not the generation of new ideas needed for long-term economic growth and environmental protection.

Overall, U.S. industry funds slightly more than one-half of the estimated $176 billion invested in R&D in this country. R&D investments vary widely among industrial sectors, ranging from 17 percent of sales in high-tech pharmaceutical areas to 0.5 percent of sales in buildings and construction. In the environmental technology industry, the percentage of R&D investment varies widely by subsector. Government investment in R&D in energy, remediation, and site assessment accounts for approximately 50 percent of the total investment. Other subsectors, however, receive nearly 80 percent of their investment in research, development, and demonstration from private funding.

Slightly more than two-thirds of the $4 billion federal investment in environmental technologies in 1994 was spent on research and development. Approximately one-half was directed to the research, development, and demonstration of pollution avoidance technologies, while the remaining investment was split among remediation, control, and monitoring technologies.

Although large firms in a number of sectors of the economy engage in R&D for environmental compliance purposes, much of the resulting technology remains internal to those firms. To date, there are few incentives for companies to cost-share this research with competitors and even fewer to license and market the resulting innovations in domestic or global markets.

A number of other factors reduce the willingness of firms, both large and small, to invest in research critical to environmental technologies. The moving regulatory horizon makes the planning of medium- and long-range research almost impossible. Uncertainties over technology performance and the market fragmentation caused by the permitting system make it difficult to anticipate the time to market a new technology or even the eventual size of the market, which adds a key uncertainty inhibiting investments. Finally, some industries have traditionally low levels of R&D spending. For those industries that are also the source of environmental problems, the government may have a role in cost-sharing some of the research that will help these industries reduce their environmental impacts.

Because of the many barriers and disincentives in the environmental technology sector and because the results of research in this area are valuable to society as a whole, the federal government plays an important role in supporting and facilitating R&D. Without this support, research of tremendous social value would not be conducted. This is particularly true for research related to environmental technologies, because these technologies not only yield profits but also provide widespread social benefits such as improved human health, environmental quality, and more habitable communities.

GOAL **Increase the overall productivity of our nation's energy, food, manufacturing, transportation, construction, and service sectors through environmental technologies and practices that significantly reduce the use of energy, materials, and other inputs. Environmental technologies are needed to better understand atmospheric, terrestrial, and aquatic systems and to remediate and restore environmental damage in a cost-effective manner.**

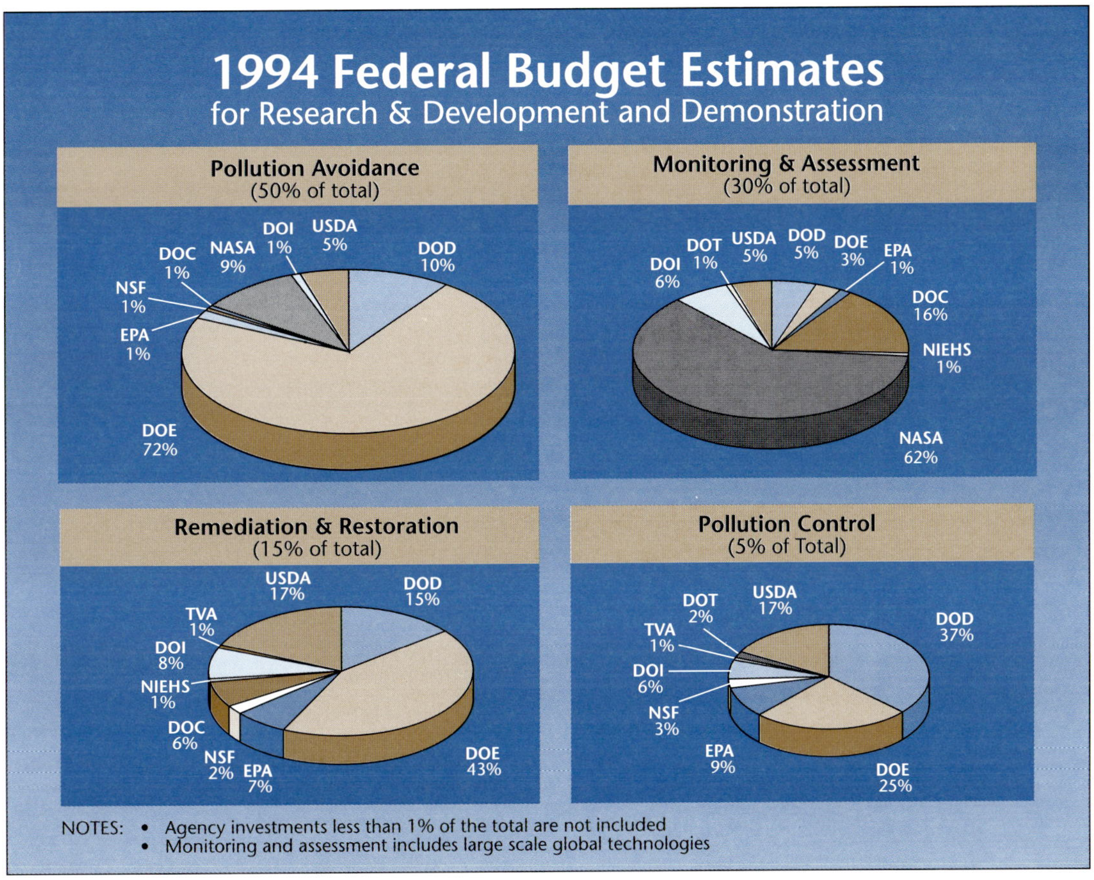

"We are spending about 20 percent of our research and development budget of a billion dollars a year on the development of new products and processes. We feel that it is not only the key to an improved environment in the future, but it is also the key to being competitive in the 21st century.... We have seen the shift to more sustainable processes resulting in some very unexpected product innovations, things that we did not anticipate when we moved into this mode."

— Robert Bringer
Vice President for the Environment, 3M

As we move from an environmental paradigm that focuses on cleanup and control to one that embraces avoidance of environmental harm, our R&D needs must likewise change. In the future, the focus will be on increasing efficiency in terms of materials and energy use. Better means of monitoring and controlling present system operations will significantly improve process efficiencies and reduce waste outputs.

Working closely with partners in industry, federal agencies involved in funding applied research relevant to environmental technologies have begun to develop new research priorities that reflect the ongoing shift toward pollution prevention and sustainability. The federal government will seek partners from the public and private sectors to pursue research and development activities in the following key areas:

INDUSTRIAL ECOLOGY: Develop a comprehensive understanding of the flows and uses of materials and energy in our industrial system and the implications of those flows with respect to the environment.

Fuel Cell R&D for Cleaner Living

Fuel cells are superefficient electrochemical devices that can convert fuels such as hydrogen, methane, or methanol from a variety of sources into electricity. They emit virtually no harmful products. The South Coast Air Quality District in California is encouraging the development of these power sources through a regulatory exemption from emissions regulations. Fuel cells can decrease energy consumption in the transportation sector, provide power, and cogenerate heat and electricity for buildings. Phosphoric acid fuel cells are nearly commercial, and even more efficient concepts are under development. In general, however, fuel cells are still too costly for major energy markets. Continued R&D is necessary to decrease the cost of components and increase performance and reliability. Fuel cells will allow increased use of biomass resources and also enable hydrogen to become the energy carrier of choice for the future.

CLEAN ENERGY: Develop clean and sustainable energy options and technologies and increase the efficiency of energy use in all major sectors of the economy.

MATERIALS, MANUFACTURING, AND COMPUTATIONAL SCIENCES: Develop science and engineering research needed to improve our understanding of the behavior of complex physical systems and design the next generation of manufacturing processes to use less energy, produce less waste, and yield easily disassembled and recyclable products.

REMEDIATION: Advance the science and technologies needed to clean up and restore contaminated sites safer, faster, and cheaper.

ECOSYSTEM RESTORATION AND MANAGEMENT: Conduct monitoring, modeling, and process research to improve understanding of threatened and damaged ecosystems, and develop technologies to restore the productivity and quality of these ecosystems.

BIOTECHNOLOGY: Enable the research and engineering of benign alternatives to current industrial products and processes.

DATA MANAGEMENT FOR DECISIONS: Provide targeted, accessible data on environmental systems and appropriate analytical tools and technologies to facilitate decisionmaking.

SUSTAINABLE COMMUNITIES: Develop integrated interdisciplinary approaches to improve the environmental, energy, and economic performance of the large technological systems needed to provide energy, transportation, housing, and other services to our nation's communities.

SUSTAINABLE AGRICULTURE: Develop technologically integrated systems for sustainable food and commodity production as well as non-polluting post-harvest food industries.

R&D PARTNERSHIPS: Leverage federal resources through cost-sharing with the private sector and various agencies.

R&D INFRASTRUCTURE: Maintain and improve the R&D infrastructure (facilities, hardware, software, demonstration sites, etc.) to support the transition from remediation and waste management to the development of avoidance technologies.

Industries of the Future: Creating Partnerships That Work

The Department of Energy, under its Industries of the Future approach, has reinvented the way that public and private R&D efforts are conducted. Under this new approach, industry sectors create scenarios that reflect the dynamic impact of key variables on their industry. The scenarios are developed by industry for industry and provide a framework for shaping major R&D needs that will position industry to be cleaner and more competitive in the next century. The result? A customer-driven research agenda that leverages scarce resources, targets public investment where there is public capacity, and assures that federal investments meet private sector needs and have high returns for society.

The Department of Energy is currently working with seven energy- and waste-intensive industries: pulp and paper, chemicals, petroleum refining, steel, aluminum, foundries, and glass. Within the manufacturing sector, these industries account for 88 percent of the energy consumed and 90 percent of the waste generated.

INITIATIVE **Within the next six months, the federal government, together with the private sector and state and local governments, will update research, development, and demonstration priorities for environmental technologies.**

Within the National Science and Technology Council, the Joint Subcommittee on Environmental Technologies enhances collaboration among federal agencies that have environmental technology programs. The subcommittee is a joint endeavor by two committees that addresses natural resources research and civilian industrial technologies. The subcommittee's strategic efforts and budget analyses are designed to increase the coordination and alignment of federal R&D on environmental technologies and the efficiency of government in facilitating R&D, demonstration, and commercialization by the private sector.

A Joint Subcommittee strategy will be reviewed by representatives from government, the private sector, and nongovernmental organizations, and will be published within six months after the issuance of this national environmental technology strategy. Action plans for the priority areas will be developed by the agencies.

DEMONSTRATION

FINDING **New approaches are needed to demonstrate innovative environmental technologies in order to answer key questions of economic viability and acceptance, expedite commercialization, and foster acceptance by stakeholders and regulators.**

Evaluations of performance and cost made during demonstrations of innovative environmental technologies enable design engineers and customers to understand the commercial usefulness of technologies. In addition, full-scale demonstrations provide opportunities for essential input by stakeholders, including the public and potential users.

Government support can expedite commercialization by providing funds to support the demonstration of technologies developed by government or the private sector. Equally important, government can facilitate demonstration of technologies, regardless of whether their development was funded by the government or the private sector, by providing sites or specialized facilities for testing, or assisting in the design of test protocols that will yield credible performance data.

GOAL **Accelerate and facilitate the demonstration of promising environmental technologies while reducing the costs.**

The federal government can encourage increased industry investment in technology demonstration through cost-shared programs and other incentives provided in partnership with the private sector. It is important for the government to facilitate systems-level demonstrations of key technologies as opposed to demonstration of single technologies. Specific projects demonstrating whole-facility and industrial ecology approaches involving business and the nonprofit private sector, including environmental groups and private cooperatives, will be crucial to gaining widespread acceptance of new ideas.

Finally, we need to increase the use of federal facilities as demonstration sites for environmental technologies developed by the government and the private sector, and establish standard performance goals and criteria for demonstrations. A number of innovative programs already exist.

STAKEHOLDER VIEWS

The existing processes governing the demonstration of innovative environmental technologies are cumbersome, consume valuable time, promote risk aversion, and continue industry's and the regulatory community's reliance on proven, but costly and perhaps less effective and efficient, environmental technology. The demonstration process for new technology must be simplified, and an attempt should be made to eliminate differing sets of regulations for individual demonstrations. Group-based support (e.g., industrial consortia) must be developed, especially in the case of large-scale demonstrations.

Through the Developing On-Site Innovative Technologies program, or DOIT, 20 western states and four federal agencies (Departments of Defense, Energy, Interior, and the Environmental Protection Agency) have formed a partnership with tribal nations, industry, and members of the affected public to develop and demonstrate new technologies for waste remediation. The partnership is based on principles of early and frequent involvement from stakeholders, timely and equal access to information, clear delineation of federal and state decisionmaking processes, flexible and realistic project schedules, and clear accountability.

Full-scale demonstrations of new approaches to *in-situ* bioremediation are being conducted at Dover Air Force Base in Delaware under a Department of Energy program. The Air Force is funding infrastructure costs, and EPA and the State of Delaware are obtaining the required permits. Once demonstrated successfully, the technologies can be used for cleanup at private sector and government sites, resulting in estimated savings of 50 to 75 percent.

These programs are important building blocks for a larger infrastructure needed to create a bridge between R&D activities and commercial applications. In the future, new programs will be needed to accommodate environmental technologies other than those dedicated to remediation and ensure wide participation of the private sector in demonstration activities.

Demonstrating New Homes

Integrated Building and Construction Solutions (IBACOS, Inc.) of Pittsburgh, Pennsylvania, is building homes that are energy efficient and environmentally responsive. The IBACOS alliance includes architects, builders, material suppliers, and manufacturers of building components, such as structural shells and systems for the kitchen, bath, indoor air quality, and home management . IBACOS, working with the Department of Energy, has moved from constructing prototype houses to planning a 25-unit community to further the research.

COMMERCIALIZATION

Domestic and global markets for environmental technologies are defined by actions of governments and the private sector. An improved relationship among these sectors can stimulate investment in technology innovation and increase the development of technologies critical to solving environmental problems. This section contains a number of recommendations designed to speed the time to market for innovative environmental technologies, provide critical capital, expand the global market, and help capture greater shares of this market.

Verifying Performance

FINDING **The lack of a system that credibly verifies the performance of innovative environmental technologies is a significant factor in limiting promising environmental solutions.**

The need for independent and credible validation of performance is almost universally acknowledged. Vendor claims for technology performance, no matter how valid, are rarely accepted by purchasers or permitters.

In the areas of pharmaceuticals, consumer products, and food, the federal government has established a standard process for assuring that performance is known and the marketplace is informed. Federal and state governments are increasingly recognizing the need for better performance verification as they see the rising cost of environmental protection and the slow rate of diffusion of efficient and less expensive solutions.

GOAL **Develop, through the federal government and the states working together, credible performance information for pre-commercial environmental technologies.**

Performance verification can significantly enhance the speed at which new technologies are commercialized and accepted by regulators, potential users, and the financial community. Verification systems must meet a number of key requirements, including credibility and accountability; flexibility, speed, and comprehensiveness of coverage; efficiency and affordability for the administering entity, the users, and the technical community; and consistency with long-term national environmental and economic objectives. In addition, the verification system must provide voluntary, universal coverage quickly and possess adequate capacity to meet demand so that the system will become an enabler for innovation rather than a bottleneck.

While federal evaluation and verification programs have concentrated on the treatment of hazardous waste sites, the need for greater focus on other areas — such as monitoring, pollution prevention, and efficient control technologies — is now clear. Ultimately, the greatest challenge for future verification programs will be to ensure the adoption of systemic, facility-wide strategies that support continuous environmental improvement, as opposed to the use of individual technologies.

The Congress recently funded a new Department of Defense program for verifying the performance of technologies that are ready for utilization in the environmental security area. DOD expects that this program will allow more than 20 additional technologies to be

STAKEHOLDER VIEWS

Even after the performance of new technologies is verified in one or more locations, they still may face the cost and time of reverifying performance as sales are sought in other companies, communities, states, or nations. The federal government should work with all interested stakeholders to design and implement responsive, cost-effective technology verification that produces readily accessible high quality data needed by those responsible for making environmental technology and permitting decisions.

demonstrated and verified each year. The program will be coordinated with EPA's new verification program described below.

INITIATIVE **The federal government will work with the private sector to establish a market-based verification process for environmental technologies. This process will be available nationally for environmental technologies within three years.**

The EPA, working with other federal agencies, the states, and the private sector, will establish a standardized verification process. EPA will seek third-party testing organizations to verify the performance of new technologies, products, and processes in all areas of environmental concern. The intent is to explore the use of a privatized verification system in order to create an efficient process by which technologies are tested and performance parameters established quickly and inexpensively. The privatization approach is being evaluated. Other options are being explored and additional feedback is being sought.

Verification systems cannot end with distribution of verified data. If the goal of verification is enabling commercialization, regional regulatory personnel and state and local authorities must be engaged and supported. "Reciprocity" will need to be arranged across governmental authorities so that verified data will be widely recognized as credible. Verified data can then facilitate and accelerate commercialization by informing regulatory and permitting decisions.

INITIATIVE **The federal government recently launched the Rapid Commercialization Initiative (RCI), which is intended to accelerate the commercialization of near-commercial environmental technologies. Over the coming year, ten technologies will be commercialized through this new program.**

The RCI focuses on helping firms maneuver past the major barriers that currently inhibit the testing and use of promising environmental technologies. The RCI is not a new spending program; rather, it is a partnership with industry focused specifically on accelerating commercialization. The RCI is designed to address a number of outstanding issues that often confound technology developers in their attempts to bring good ideas to market. These include three major barriers: a lack of public and private test sites to demonstrate technologies at full scale, the lack of credible verification procedures, and the inability to obtain testing permits quickly.

A competitive process will be used to select participating companies. RCI will provide them with support and help in finding test sites for full-scale demonstrations, developing a plan for demonstration and performance tests, obtaining environmental permits needed during the demonstration and testing phases, conducting the testing, verifying the results, obtaining the initial operating permits, and disseminating the results of performance testing. Experience gained from the RCI projects will be used in ongoing efforts to design and implement a larger commercialization process.

The RCI will be administered through the Interagency Environmental Technologies Office with support from a number of federal agencies. A Notice of Public Intent describing RCI is scheduled to be published in the *Federal Register* in the summer of 1995.

"The nation's civil engineer/environmental professionals represent an industry that contributes more than 10 million jobs and 13 percent of GDP to the nation's economy. We propose a national environmental 'partnership,' focused on the accelerated introduction of those technologies that will both resolve long-standing environmental problems and prevent future environmental degradation."

— Civil Engineering Research Foundation

An Innovative Approach to Verification in California

In early 1994, the State of California EPA's Department of Toxic Substances Control initiated a "Certification Program for Hazardous Waste Technologies" under authority of Assembly Bill 2060. More recently, the state established authority for a similar process for the technologies of air pollution management. The California certification program is a model for verification and certification processes that could be emulated in other states and at the national level. One of the first companies to use the program saw its sales increase by six times since its product received CalEPA certification.

Financing

FINDING **Financial uncertainty and a high level of risk limit the availability of investment capital for environmental technologies.**

Developing and commercializing environmental technologies requires capital, as does starting and expanding the companies that deliver environmental products, processes, and services to domestic and international markets.

Although the environmental technology industry is larger than many other sectors of our economy, it attracts very little private capital. In 1993, for instance, studies indicate that approximately $31 million in venture capital was invested in conventional control and remediation technologies, supporting just 12 firms. In 1994, this amount dropped to $25 million invested in fewer than 10 companies.

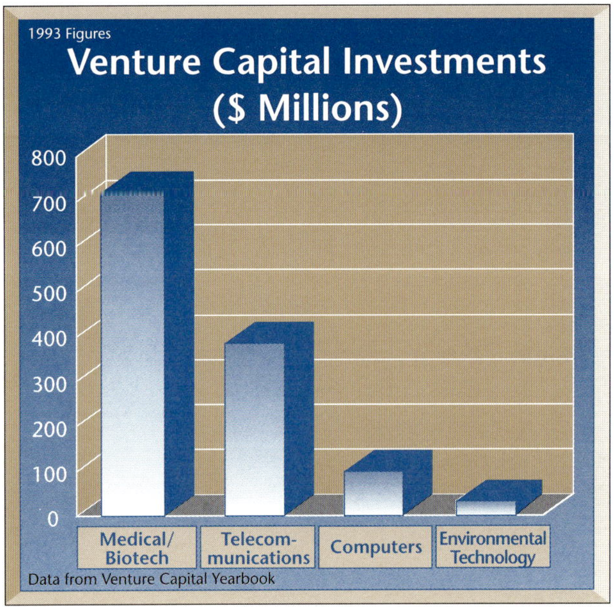

A number of reasons account for this industry's tendency to repel capital. Government environmental policies and regulations are important drivers of the market, but the timing and size of current and future markets is often a function of the specifics of regulation, including the timetable for new regulations, the stringency of current standards, and their enforcement.

The Export-Import Bank

The mission of the Export-Import Bank, which was created in 1934 as an independent U.S. government agency, is to create jobs through exports. The bank does not compete with private lenders but accepts risks they will not undertake. The bank provides insurance coverage of commercial bank loans made to foreign buyers of U.S. goods and services, credit insurance to protect businesses against political and commercial risks, and a capital guarantee program to support export-related production and marketing activities. The bank has made the development of environmental export a high priority and has launched a new Environmental Exports Program, which will provide enhanced support to exporters of environmental goods and services and a new insurance policy for small business exporters.

The unintentional result is to increase the financial uncertainty and risks associated with investing in environmental technologies. This uncertainty exists for providers of venture capital, as well as other potential investors such as banks, and for many insurance companies, which often underwrite significant environmental risks of client companies.

Exporters face additional financing challenges. All companies, but especially the small and medium firms that make up the majority of the industry, have difficulties finding providers of capital to finance a move into foreign markets. Such firms find it difficult to compete for export opportunities that have long lead times and require considerable financial and legal sophistication — including dealing with multilateral development banks.

Finally, in many developing countries, government funds for environmental protection are scarce, and likely to remain scarce in the foreseeable future. The lack of adequate financing not only remains a special problem for all environmental companies but also will hinder environmental protection worldwide.

GOAL **Ensure that adequate investment capital is available for the development, commercialization, and use of environmental technologies, both for entry into the domestic market and to support an active export industry.**

Domestically, the federal government is largely an enabler, rather than a provider, of capital. Through its policies and actions, however, it can change the rules of the game and significantly lower the risks for private sector financiers.

The federal government can help increase the predictability of environmental technology investments through reforms of the environmental regulatory system. Firms need greater flexibility in managing for environmental results to reduce the uncertainty associated with existing environmental regulations.

> *"Generally, capital is a follower, not a leader. ... If we cannot make money, we will not come to the table. We have to create an industry where we can demonstrate that environmental technologies can make money."*
>
> *— Frank Pope*
> *Technology Funding, Inc.*

The federal government will support efforts to reduce the fragmentation of the market for environmental technologies. Current activities at a state and regional level for facilitating the exchange of performance data for environmental technologies and ensuring reciprocity of approvals are critical in this regard.

In addition, the federal government will work with the states to reduce the time to reach market by streamlining the permitting process and accelerating technology verifications.

Finally, the federal government will work to improve data on existing and future needs and markets for technologies domestically and globally.

Internationally, the Department of Commerce's Environmental Trade Working Group is helping businesses of all sizes gain access to vital financing for feasibility studies, overseas sales, and investment through programs of the Export-Import Bank, Trade and Development Agency, Overseas Private Investment Corporation, and Small Business Administration.

Promoting Exports and Market Development

Although the United States is widely regarded as the world leader in design and construction services, pollution monitoring and assessment, solid and hazardous waste management, and remediation of contaminated sites, U.S. international leadership cannot be taken for granted.

FINDING **Due to tremendous domestic demand and the limited financial strength and export experience of small- and medium-sized companies, the U.S. environmental industry has mostly focused on the domestic market. Many U.S. environmental companies have difficulty competing internationally, resulting in missed opportunities to support or create new U.S. jobs.**

The tepid export performance of U.S. environmental technology firms stands in marked contrast to the export-oriented industries in Germany, Japan, France, and the United Kingdom. The environmental industries of these nations are mounting a competitive challenge with active government support that combines political, technological, financial, and commercial resources. Japanese companies, for example, are encouraged to develop and commercialize energy-efficient and environmental technologies through a wide range of governmental incentives. Japan has committed more than $2.2 billion over 10 years to finance the export of environmental technologies and services. Germany is moving aggressively to apply environmental technologies domestically and to capture a larger share of global markets. British and French companies dominate the international markets for water and wastewater treatment technologies.

The United States, on the other hand, lags significantly behind its competitors in committing public resources to trade promotion. Among the world's major trading nations, the United States ranks last or next to last in export promotion indicators such as spending per capita, spending as a fraction of gross domestic product, and spending as a fraction of manufacturing exports.

GOAL **Increase U.S. environmental technologies exports to support and create new high-paying U.S. jobs and to contribute to the achievement of sustainable development.**

To compete for export markets and remain internationally competitive, governments and industry in the U.S. must work together to advance the commercialization of environmental technologies. Supporting environmental technology exports not only helps build a foundation for sustainable development abroad but also creates or supports jobs in the United States.

President Clinton, in his Earth Day 1993 speech, directed the Department of Commerce, Environmental Protection Agency, and other key agencies to assess the domestic and international competitiveness of the U.S. environmental technologies industry and to formulate a strategy that gives U.S. companies the assistance they need to increase exports and create U.S. jobs. Responding to the President's mandate, the Administration released its Environmental Technologies Export Strategy in November 1993.

> *"Over the long term, improvements in environmental technology, particularly when it comes to pollution prevention, are going to be critical to the ability of American companies to compete. Not only do new technologies reduce compliance costs, they improve competitiveness because they lead to greater efficiency."*
>
> — U.S. Senator Joseph Lieberman

The export strategy is designed to enhance the competitiveness of U.S. firms, particularly small- and medium-sized enterprises. It rests on strong public-private partnerships and streamlining cooperation among U.S. government agencies. The 35-member Environmental Technology Trade Advisory Committee was created in December 1994 to strengthen the direct link between the government and the environmental technology private sector. The committee has already begun to advise on implementation of a set of coordinated, targeted actions.

The export strategy ensures that U.S. firms receive full advocacy support from their government when seeking contracts overseas. Under this strategy, an Advocacy Network of agencies is helping the U.S. environmental industry win bids for major projects and commercial transactions by providing government support to assist the foreign decisionmaker in selecting U.S. companies. Since the creation of the Advocacy center in the Department of Commerce in 1993, the administration has successfully advocated for more than $2.5 billion in environmental technologies exports.

The export strategy focuses export promotion efforts on high-growth markets that offer the best opportunities for U.S. firms. In 1994 the federal government identified five priority markets or regions for environmental exports: Mexico, Chile and Argentina; South Korea; Poland; the Czech Republic; and the China Economic Area. Concise export market plans have been produced for each of these regions, and India, Brazil, and Turkey will be added in the immediate future. A number of state and local governments are also developing regional coalitions and strategies to help their environmental technology industries move into high-growth markets.

The export strategy recognizes the important role U.S. bilateral and multilateral negotiations play in eliminating or substantially reducing non-market barriers that inhibit the penetration of foreign markets by U.S. environmental technologies. Under this strategy, the federal government and its private sector partners will continue to support vital U.S. international negotiations and participate in international activities to set standards to ensure that existing export markets remain open to U.S. environmental technologies and that new markets can be entered.

Finally, the export strategy calls for establishing better communication between foreign buyers and American sellers of environmental goods and services. Communication between American technology developers and sellers and the federal government can alert the U.S. industry to trade leads. The federal and state governments as well as nongovernmental organizations have an important role to play in providing this type of information.

Although the federal government will continue to focus implementation of the Environmental Technologies Export Strategy on key emerging markets, the activities

ENVIROTECH-ON-LINE

The International Environmental Business and Technology Institute, Inc. and the University of Massachusetts, in partnership with the Department of Commerce, have developed ENVIROTECH-ON-LINE, a Global Environmental Business and Technology Data System that helps U.S. firms obtain the information they need to expand their environmental technology exports.

The system is designed to increase access to strategic information about environmental markets, financing, and contacts. Users can easily search this information using plain English sentences with INQUERY, an advanced information retrieval system developed at the National Science Foundation's Center for Intelligent Information Retrieval. Users access the information system and use INQUERY through the Internet or by CD-ROM.

initiated in 1995 will extend our cooperation with the private sector to commercialize and diffuse U. S. environmental technologies worldwide.

Together with the private sector, the Administration is demonstrating U.S. environmental technologies and promoting these technologies in new markets. In addition to demonstrating environmental technologies, the Administration also seeks to foster innovative financing that will make commercialization of new technologies feasible in selected markets.

In December 1994, during the Summit of the Americas, countries of the western hemisphere initiated a dialogue on creating partnerships to strengthen our hemisphere's capacity to prevent and control pollution and use natural resources on a sustainable basis. These regional partnerships will provide the management capability to ensure that nonenvironmental business proposals are "packaged" with the necessary environmental technologies and services.

FINDING **Sustainable development and the dissemination of environmentally beneficial technologies are hindered by inadequate infrastructure, human and institutional capacity, and policies and regulations in developing countries. Investments are critically needed to create an international environment that enables the diffusion of environmental technology.**

Foreign assistance plays a pivotal role in creating demand for environmental goods and services. It will foster environmental stewardship while facilitating the dissemination of technologies globally. The United States must invest in developing countries to build the proper "enabling environment" for environmental technologies. By working with nations to facilitate environmentally sound economic development, the United States promotes environmental stewardship and builds markets for environmental technologies. A great need exists for stronger international development programs by individual nations, including the United States, and multilateral organizations.

GOAL **Build a foundation for environmental stewardship and sustainable development internationally by implementing a coordinated set of activities to facilitate avoidance of environmental harm and remediation through the development, adaptation, and use of environmental technologies.**

Throughout the world, the private, public, and nonprofit sectors are helping developing countries build the foundations for sustainable development and the widespread use of environmental technologies. Assistance programs can facilitate sound environmental management and create an enabling environment for sustainable economic growth in developing countries.

Information dissemination and exchange are essential building blocks for the development of local capacity to address environmental challenges. The U.S. Agency for International Development's (USAID) Global Energy and Environment Network (GLEEN), for example, is currently working in the Philippines to create a locally appropriate electronic communications infrastructure to deliver information on environmental technologies and foster discussions on energy and environmental problems. Targets of an upcoming expansion of GLEEN are India, Mexico, Indonesia, and Brazil. The Coalition for International Environmental Research and Assistance, a nonprofit network of environmental experts, is working with the governments of developing countries to establish environmental priorities and possible program initiatives.

Training and education build the capacity to develop, assess, and adapt technologies. They also create the human resource base to design, institute, and enforce policies and regulations.

Policy and regulatory reform are critical to creating the proper incentives for environmental stewardship and environmental technology markets. Trade regulations, standards, codes for buildings and industrial facilities, and pollution laws all create or dilute demand for particular environmental technologies, goods, and services.

Adequate financing mechanisms are essential to the introduction of new environmental technologies. For example, the Environmental Enterprises Assistance Fund, supported by USAID and a number of foundations, promotes commercial technologies that yield environmental benefits by providing risk capital and management assistance to build in-country capital markets while demonstrating the potential value of environmental investments. Foreign aid can also be used to help locate, leverage, and direct funding by others. In Costa Rica, for example, USAID technical assistance is helping support implementation of an Inter-American Development Bank loan to fund the introduction of energy-efficient improvements with the national utility.

Demonstration and pilot projects provide empirical proof of the viability of a technology in a new market. In Tunisia the introduction of improved technologies

STAKEHOLDER VIEWS

Better coordination is needed among federal agencies engaged in initiatives to help develop international markets for U.S. environmental technologies. International training and institutional capacity-building are fundamental building blocks for technology diffusion and viable markets. Well-trained professionals in potential client nations are necessary to build the institutional capacity and economic and regulatory infrastructure required for future markets. Public-private partnerships and collaboration are also key.

"States should cooperate to strengthen endogenous capacity building for sustainable development by improving scientific understanding through exchanges of scientific and technological knowledge, and by enhancing the development, diffusion, and transfer of technologies, including new and innovative technologies."

— Agenda 21

UN Conference on the Environment and Development, 1992

saved a battery firm $1.5 million annually, reduced lead seepage into groundwater, and reduced workers' exposure to lead dust. In Thailand USAID, the Royal Thai Government, and the U.S.-Thai Business Partnership are developing and demonstrating two- and three-wheeled electric vehicles for commercialization. These vehicles decrease the harmful emissions, noise pollution, and lead-related health risks associated with current Asian transportation systems.

The USAID-led U.S.-Asia Environmental Partnership is using business exchanges, information dissemination, training and small grants to promote partnerships to solve environmental problems.

Under EPA's U.S. Technology for International Environmental Solutions program, the EPA and the U.S. Department of Agriculture are working with American vendors, universities, and Mexican authorities to demonstrate the performance of low-cost, reliable, and easy-to-operate water treatment plants for three small Mexican communities. The demonstrations will promote the application of proven and cost-effective American technologies to solve a pressing environmental problem.

INITIATIVE **The U.S. Agency for International Development, as part of its Initiative for Environmental Technologies will work in partnership with the private sector to focus development assistance on addressing critical environmental challenges in developing countries.**

The Initiative for Environmental Technologies will help create markets for environmental technologies and eliminate barriers to the flow of technology through partnerships with the private sector. This initiative will emphasize international collaboration with small- and medium-sized U.S. environmental technology firms.

The Initiative for Environmental Technologies will help small businesses abroad come into compliance with newly strengthened environmental regulations, conduct industrial audits to identify opportunities to prevent pollution and conserve energy, and promote development of environmental enterprises. The initiative will engage the expertise of U.S. environmental businesses in undertaking audits and training, and promote joint ventures between U.S. and foreign manufacturers of environmental technology.

This initiative will focus on key needs for policy and regulatory reform, as well as the development of financing mechanisms and training. Additionally, the initiative will be used to disseminate information on environmental problems abroad and foster specific problem-solving discussions. The initiative will serve to coordinate U.S. development assistance in energy technologies, urban infrastructures, and industrial pollution prevention on a global basis.

The Presidio: A Laboratory for Sustainable Practices

The Presidio of San Francisco, a new addition to the National Park system, is becoming a working laboratory where concepts of environmental sustainability are shaped and formed. Public and private organizations are converging on the Presidio to collaborate on finding solutions to environmental problems through research, demonstration projects, education and training, and community service programs. Visitors to the Presidio will have an opportunity to view and learn about practical approaches to achieving sustainability.

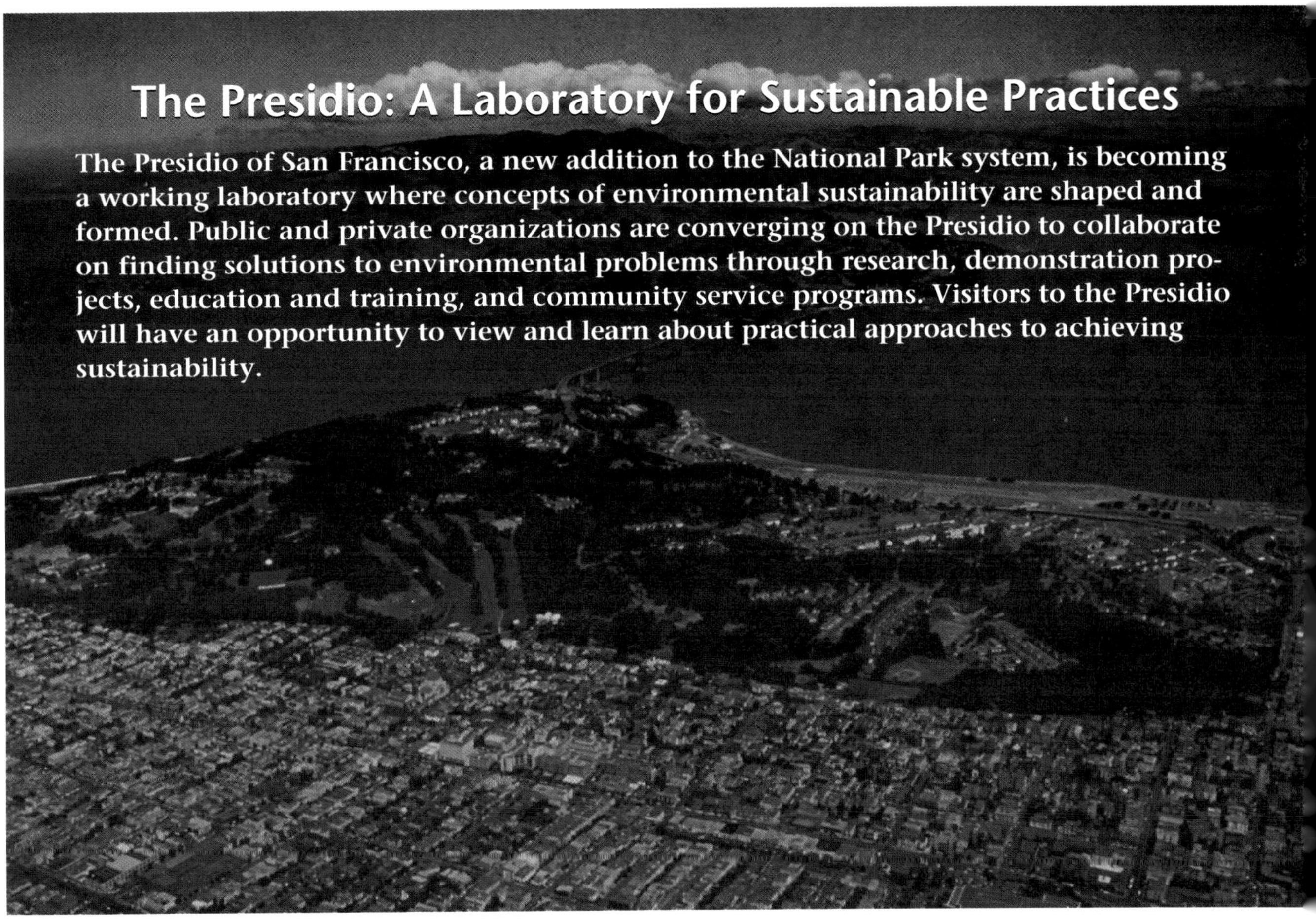

SUSTAINABLE COMMUNITIES

FINDING **Our nation's future strength will in large part be built on the viability of our nation's communities. We must make choices today that increase the sustainability and desirability of our cities, towns, and rural areas if we are to preserve our natural environment and build a strong domestic economy.**

Many environmental technologies will ultimately be used in local communities, schools, businesses, and homes. Environmental technologies can help existing and developing communities — urban, suburban, and rural — become more sustainable.

STAKEHOLDER VIEWS

To achieve sustainable development, local communities must be actively involved in the choice and application of environmental technologies. Achieving sustainable communities requires understanding the interdependence of technological systems and integrating these systems in cities and rural areas. An interdisciplinary approach is needed that considers major aspects of technology decisions in social and environmental contexts, and allows for an interactive relationship to be developed between the community and the developers of technologies. Local involvement is critical.

Communities are complex and interlocking environmental, social, and economic systems. We need a clear understanding of the preferred future for these systems and better cooperative planning to achieve that future. A critical part of that strategic planning is citizen involvement to ensure that technological choices support the goals and aspirations of the community. It will also require a broad approach that allows individuals and organizations to cross traditional boundaries and work together to integrate economic, environmental, and social policies. Our environmental technology strategy stresses communications, access to information, resource management, and democratic participation in the decisions which will shape the technological future of communities.

GOAL **Develop and implement sustainability plans in many U.S. communities and make significant progress toward achieving sustainable communities over the next 25 years, increasing the quality of urban, suburban, and rural life and reducing our use of energy and natural resources.**

To increase the compatibility of the built and natural environment, we will need an improved understanding of the environmental, social, and economic trade-offs involved in the choices facing our cities and towns. The overarching technological challenge is to recognize that technology is both the means to a better society and a force with potentially unintended consequences.

Energy, water, transportation, and building infrastructures both enable and constrain development. Since investments in infrastructure last for decades, or even centuries, technological choices must be made carefully. Integrating technology development with the concepts and practices of ecosystem management, rural and urban development, and an understanding of the unique cultural, biological, and social circumstances of each community, are important for the development of sustainable communities.

Some communities and businesses have already started on the path to sustainability. Other communities can begin programs oriented toward sustainability with state, local, and federal governments; nongovernmental organizations; universities and colleges; businesses and local citizens. For instance, Seattle has explicitly incorporated sustainability concepts in its community planning and development to help improve the quality of life and the environment. At Clagett Farm in Maryland, the Chesapeake Bay Foundation, other nonprofit organizations, the agricultural community, University of Maryland, Environmental Protection Agency, and U.S. Department of Agriculture are developing and implementing new sustainable agricul-

Chattanooga: Moving Toward Sustainability

In response to the dual problems of inner city decline and severe environmental degradation, Chattanooga has incorporated sustainable community concepts into the city's planning for urban development. In 1984 Chattanooga developed a community-wide vision that identifies 40 specific goals. Reducing dependence on automobiles through reliance on clean public transportation (electric buses), strategically located multi-purpose facilities, and the integration of affordable housing with workplaces and entertainment sites will reduce air pollution and improve the quality of life for the region's residents.

The city is working with the Department of Energy's facilities at Oak Ridge to develop and commercialize environmental technologies that will assist businesses in conserving energy, minimizing waste streams, and maximizing recycling of products and effluents. These technologies will promote evolution to zero emissions manufacturing. As part of this overall process, the Tennessee Valley Authority is balancing the advantages and disadvantages of various energy sources such as hydroelectric, gas, coal, nuclear, and biomass fuels in an integrated approach to regional growth and development.

tural practices to reduce impacts on the bay while increasing farm income. In the private sector, Stemilt Growers Association, representing more than 250 fruit producers in Washington state, has voluntarily achieved over the past six years a 70 percent reduction in post-harvest chemical use while improving fruit quality.

INITIATIVE **Communities will better integrate the development and use of environmental technologies into sustainability plans and activities.**

Cities, towns, and other local governments will continue to develop and implement sustainability plans in their communities in partnership with businesses, universities, colleges, nongovernmental organizations, citizens, and the federal government. Individual citizens factoring environmental considerations into their daily lives through their choices in purchasing, commuting, and other actions can help further the development of sustainable communities.

A special conference focusing on environmental technologies in sustainable communities should be convened as a joint activity of the National Science and Technology Council, the President's Council on Sustainable Development, and communities leading sustainability efforts. The Council on Sustainable Development is developing and recommending to the President a national sustainable development action strategy.

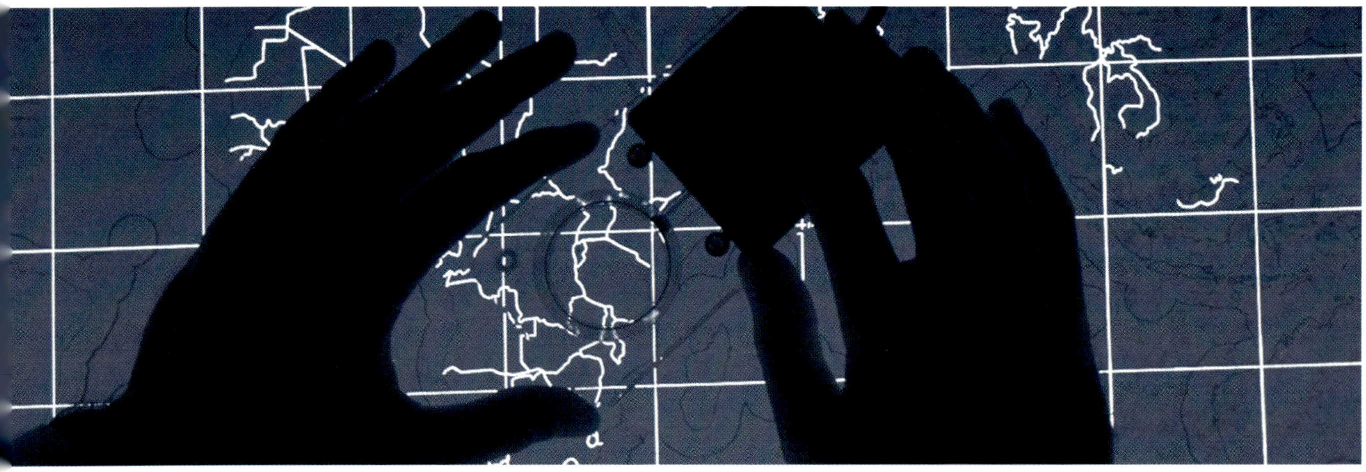

Learning and Working Together

This theme recognizes that the generation of knowledge and its transmission and sharing are our greatest assets in the development of novel ideas and their transformation into environmentally and economically useful technologies. The key elements of learning and working together — coordination, education, and information — establish a framework for defining new policies and strategies to advance environmental technologies. Together, these elements create the knowledge base and define the actions that are central to a participatory approach to technology development and diffusion. They enable individuals, businesses, and communities to learn from and communicate with each other as they make daily decisions related to purchasing, commuting, manufacturing, and planning.

COORDINATING ACTION

FINDING **Domestic and global markets for environmental technologies are influenced by actions of governments and the private sector. An improved relationship among all levels of government and the private sector can stimulate investment in environmental technologies and increase their diffusion.**

When governmental policies inhibit innovation, lead to an imbalance in competition, or unduly limit the flow of financial resources, the growth of national economies slows and environmental protection is threatened. At the recent White House Conference on Environmental Technologies, the Vice President announced that the federal government will be more accessible to the private sector and the public and will take steps to increase collaboration among federal agencies so that government is more efficient and works more effectively with the private sector, the states, municipalities, and others.

STAKEHOLDER VIEWS

Businesses need access to government information on technologies, technology applications, financing, and regulations. Unfortunately, many firms, especially small companies, do not have the time and resources to confront the government's "information jungle," complex organizational structures, and other roadblocks. The government should create a "one-stop-shop," including an Internet-accessible database of commercially available technologies, private and public sector needs, international market opportunities, financing sources, and business startup assistance.

GOAL **Ensure that the federal government becomes a more accessible partner with the private sector in advancing the development of environmental technologies.**

Critics of federal programs indicate that it is often difficult to identify programs within federal agencies and access government information and assistance. Many opportunities exist to increase the effectiveness of federal programs through collaboration among agencies and to facilitate the process of partnering between government and industry.

Progress is being made. For example, the Departments of Energy and Defense and the Environmental Protection Agency are working together with private organizations to facilitate use of government-owned locations for testing and demonstration of environmental technologies. DOE and DOD are also coordinating performance testing and are working with the states to increase the interstate recognition of performance data developed at federal sites. All of these efforts will facilitate commercial introduction of innovative solutions.

GNET: Enhancing Communications About Environmental Technology

The Global Network for Environmental Technology (GNET) is a user-friendly electronic communications system for exchanging information and locating opportunities in the environmental technology arena. The GNET system includes computer-based project management capabilities, groupwork systems, news services, and full access to the Internet and World Wide Web. GNET provides access to federal and other environmental technology databases with detailed information on technologies, testing facilities, domestic and international market opportunities, regulatory policies, and funding sources.

In the future, companies will have easier access through a single point of contact to information about federal policies, regulations, and programs that are related to environmental technologies. An internal point of coordination will link federal agencies active in fostering research and development, commercialization, and use of environmental technologies. Two principles will guide these efforts: a private sector focus and a comprehensive approach embracing all types of environmental solutions, including avoidance, monitoring, control, remediation, and restoration.

INITIATIVE **The federal Interagency Environmental Technologies Office will act as an information clearinghouse and bridge between the federal government and the private sector in advancing environmental technologies.**

Responding to stakeholder views regarding the necessity of obtaining government information on environmental technologies, financing, and regulations, the new Interagency Environmental Technologies Office (IETO) will provide a location for interagency activities and a point of contact for the private sector and state and local governments. IETO will also promote intra- and interagency cooperation in environmental technology activities such as implementation of the Rapid Commercialization Initiative. IETO will facilitate information sharing on environmental technologies among federal, state, and local governments, industries, and educational organizations. IETO will develop a one-stop information clearinghouse and utilize existing private sector and government capabilities to disseminate information on environmental technologies.

EDUCATION AND TRAINING

FINDING **Education and training programs at all levels, from elementary schools to universities, have not sufficiently integrated information on the challenges and opportunities associated with sustainable development and the role of environmental technologies in achieving it.**

Advancing the development and application of environmental technologies over the next 25 years will depend to a large degree on public awareness of environmental challenges facing the nation and the world and opportunities to address them. Collectively, the citizens of the United States are the strongest force for advancing environmental technologies because decisions made by individuals at home and in the workplace will shape markets and largely determine when and how these technologies will be applied.

Citizens must have the basic knowledge necessary to make reasoned judgments about environmental risks and ways to reduce them. This is particularly true if regulatory reform and an increased emphasis on partnerships gives local communities more influence over regulatory decisions and the allocation of public funding to support their development. Perhaps most importantly, individuals must be aware of the impact of their own activities on their local environments if society is to achieve sustainable development.

Expanding the application of environmental technologies will depend to a large degree on the extent to which business leaders and engineers understand the economic opportunities associated with pollution avoidance. Consequently, the nation's business and engineering schools, as well as other educational institutions, must direct more attention to environmental issues and the appropriate application of environmental technologies. Operating the manufacturing plant of the future also will require a new cadre of well-trained environmental technicians and professionals. Consequently, the community college system will play a critical role in education and training related to environmental technologies. Emphasis on the training and retraining of professionals is crucial.

STAKEHOLDER VIEWS

Education about the environment is essential if sustainable development is to be achieved. In general, the public is insufficiently informed to participate effectively in the decisionmaking process regarding the development and deployment of advanced technologies. The federal government needs to serve as a catalyst in developing a comprehensive approach to environmental education and training, encompassing education of students at all levels, employee training, retraining of displaced workers, youth-at-risk programs, and teacher training. Consumers need to become increasingly aware of the environmental implications of their purchases and behavior.

The Department of Labor (DOL) will collaborate with the Department of Energy to examine ways to involve workers in improving energy efficiency and reducing waste, and will include the results in the New DOL Occupational Information Network. This system will allow the Labor Department to assist industry with skill identification and acquaint workers with the expertise necessary for the development and use of environmental technologies.

GOAL **Build an integrated, interdisciplinary environmental education and training system for students at all levels over the next decade.**

Environmental problems from polluted air and water to climate change, stratospheric ozone depletion, and biodiversity loss cut across a range of ecological and geographic regions and have broad impacts on society — from public health to economic growth. Studying these cross-cutting problems and responding to them necessarily involves a number of disciplines in the natural and social sciences. Accordingly, educational institutions at all levels must work to integrate environmental considerations into their curricula to help educate and train the next generation of leaders who will face these environmental challenges.

"Think globally and act locally."

— Rene Dubois

The present departmentalization of disciplines in educational institutions makes teaching about the environment and technology a particular challenge. During the past several years, considerable progress has been made in integrating environmental concerns into the curricula of the nation's universities and community colleges. Universities and community colleges must continue to advance environmental education programs by developing new teaching materials and employing integrated multidisciplinary approaches to the development of curricula. Courses should include the basic elements of environmental science, engineering, and economics to ensure a more informed public and skilled workforce. To assist K 12 schools in developing and offering programs that incorporate environmental concerns in a number of disciplines, federal grant programs will be encouraged to provide greater incentives for adopting multidisciplinary approaches to problem solving.

In the years ahead, environmental education programs must be integrated with other community activities such as training youth at risk, fostering environmental justice, and retraining displaced workers. Already, in many cities throughout the United States, inner city youth participate in community environmental programs and learn the importance of the environment in the context of their own

Global Environmental Protection: A Role for the World's Children

On Earth Day 1994, Vice President Gore announced the Global Learning and Observations to Benefit the Environment (GLOBE) program. GLOBE is an international environmental education and science effort designed to enable students, educators, and scientists to work together to monitor the global environment and provide information for developing the first worldwide environmental database.

The GLOBE program, with participating schools around the world, will allow students to perform environmental measurements that will greatly augment earth observations from existing satellite and ground-based systems. Scientists and educators are working together to design experiments that will provide hands-on science and math experience for elementary through high school students and generate useful environmental data for scientists. Students participating in GLOBE in the United States and around the world will use state-of-the-art technology to monitor their local environment, including weather, air and water quality, and biological measurements, and then be able to view their data in a global perspective and relate it to other environmental indicators.

communities. Also, a number of defense firms have begun retraining some of their personnel in environmental engineering and environmental technologies to provide them with new job opportunities.

Through an unprecedented collaboration, the Departments of Education and Labor created the School to Work program to help prepare the nation's workforce for the challenges ahead as we move into the 21st century. School to Work is based on the notion of partnerships involving schools, businesses, and labor; states and the federal government; parents and teachers; and students and employers. The growth of job opportunities in environmental technology necessitates that environmental education become an integral component of programs such as School to Work.

"I believe that being an institutional leader in the advancement of sustainable development will give our university a competitive advantage in today's educational, research, and health care marketplace. To put it simply, we view this initiative as just plain good business."

— Stephen J. Trachtenberg, President
The George Washington University

Environmental education must go well beyond the nation's schools to daily activities. Existing organizations such as 4-H, which has local chapters in every county of the country, can be used to advance community-based education about sustainable development, pollution prevention, and the appropriate application of environmental technologies. Federal and state agencies should identify existing educational programs and work to incorporate environmental considerations into them.

INITIATIVE **Federal agencies will work collaboratively with other organizations to foster and promote a new ethic within the system of higher education.**

The objectives outlined here will be accomplished through a newly formed public-private partnership and other to-be-developed pilot programs directed toward our colleges and universities and those of other nations. In December 1994, EPA and the George Washington University in Washington, D.C., announced a unique partnership to develop the nation's first model "green" university. The university has committed to incorporating an environmental ethic and the principles of sustainable development in all of its activities. This commitment encompasses its education and training programs, research projects, and university services such as health care. The university and several national and international organizations are in the process of establishing a National Environmental Resources Center to provide user-friendly computer access to information on the environment, energy, green buildings, and sustainability.

INITIATIVE **The federal government will work with individuals from industry, states, and municipalities, nongovernmental organizations, and the educational community to develop a blueprint for building successful partnerships to support environmental education and training efforts.**

In the fall of 1994, leaders from education, business, government, and non-governmental organizations around the country met at the Presidio in San Francisco, California, to explore ways to build effective partnerships to support environmental education and training activities. At the National Forum on Partnerships Supporting Education about the Environment, more than 100 individuals with a broad range of expertise discussed their individual and collective roles, reasons for forming partnerships, common visions, and opportunities for partnership. The participants outlined plans for action, which are being developed with the help of others in areas such as global information access, integration of sustainable development principles into education, and resource coordination.

INITIATIVE **Federal agencies will work closely with the nation's network of community colleges to provide information, technical resources, curricula, and, where appropriate, funding to advance educational and training programs related to sustainable development and environmental technologies.**

The community college system is the fastest-growing sector of educational institutions in the United States. The nation's 1,200 community colleges are well connected to businesses in their localities, making them ideally suited to advance environmental education and training programs. In addition, community colleges have a proven track record in worker retraining, and they can easily modify their curricula to meet the evolving needs of industry and their local communities.

The Department of Labor, using information provided by the Department of Energy, will help the American Association of Community Colleges provide the community college system with model curricula to train workers, especially in small- and medium-sized firms, about energy efficiency and waste management techniques.

The Department of Labor will also use its Training Technology Resource Center to disseminate information on the Department of Energy programs and other federal, state, corporate, and organized labor programs that provide skills needed for environmental cleanup and waste reduction. Further, the Labor Department will explore, with the Energy Department, ways to introduce energy efficiency and waste-reduction skills into job training programs such as School to Work and the Job Corps.

The Partnership for Environmental Technology Education (PETE) is a national nonprofit organization designed to link participating community colleges with the technical resources found in federal and state agencies, private industry, and professional societies. The PETE program for environmental technicians helps develop curricula designed to encourage students to pursue degrees in environmental science, engineering, and management at four-year institutions. Through its network of six regional public-private partnerships serving all parts of the nation, PETE also works to enhance the participation of traditionally under-represented groups in environmental fields.

The Interagency Environmental Technologies Office will serve as the focal point for connecting federal environmental technology programs with the community college system. The interagency office will support and work to expand successful programs such as PETE. In the next year, the office will build on existing interactions with the community colleges and establish information-sharing and outreach mechanisms.

MONITORING, ASSESSMENT, AND ENVIRONMENTAL INFORMATION SYSTEMS

Quality information, broadly accessible to all decisionmakers, is vital to achieving the vision outlined earlier in this strategy document. In working toward sustainable development, decisionmakers must have ready access to integrated economic, environmental, and human health-related data. Businesses, for example, need access to information on material flows, engineering design, and life-cycle assessments to apply industrial ecology concepts effectively. Industry and governments require timely and accurate information on the occurrence, intensity, and potential impacts of natural hazards and climate change to guide investments and make operating decisions. And environmental technology developers, producers, and financiers need timely and comprehensive information on markets and technology needs.

FINDING **The considerable national investment in environmental monitoring, assessment, and related information systems made by all levels of government and the private sector should be fully and cost-effectively linked to support efforts to achieve sustainable development.**

Cost-effective monitoring, assessment, and information systems are required for societal and economic decisions to be coupled with a comprehensive understanding of the environment. Describing and predicting changes in the Earth's environment, and conserving and managing wisely the nation's natural resources to ensure sustainable economic opportunities, depend on highly leveraged monitoring and information dissemination partnerships.

Although substantial progress has been made over the past 25 years, data generated, assessed, and disseminated with the help of advanced technologies are needed to identify and assess emerging environmental problems locally, regionally, and globally. Currently, policymakers encounter a wide range of challenges in accessing and using environmental information in decisions. Many lack adequate or timely information for assessing environmental alternatives and consequences. They also lack the tools required to relate everyday decisions to impacts on long-term sustainability.

As more communities begin to work toward sustainability, demand for environmental information will increase. For example, effective urban planning requires reliable information about the environmental impacts of proposed development, and sound ecosystem management requires reliable information about interactions among soils, water, flora and fauna, and climate. In general, public information about environmental monitor-

STAKEHOLDER VIEWS

Approaches to the acquisition, access, and use of environmental data for regional and local applications raise many policy, procedural, and legal issues. Valid procedures must be established to foster trust and reciprocity in sharing and interpreting environmental information. The nation must simplify and standardize information system linkages among data sources, including the federal government and local decisionmakers such as planning commissions. Improvements are needed in the areas of data ownership and maintenance responsibilities; data quality standards and validation; and data processing, modeling, and applications techniques.

ing and assessment capabilities, databases, and technologies is not readily available to a broad array of organizations addressing environmental concerns. If the nation is to achieve sustainability, the federal government working closely with regional and local agencies must expand efforts to improve awareness of environmental information resources and access to them.

To improve our ability to predict environmental change, the nation must implement an integrated global observing capability that effectively combines both satellite and ground based measurements. At the same time, we must ensure that data are processed, analyzed, and distributed in a timely fashion. The nation must continue to invest in these capabilities, which serve as the supporting foundation for decisionmaking. We must guard against technological obsolescence.

GOAL **Improve the nation's environmental monitoring data and information systems substantially over the next five years through public-private partnerships designed to maximize opportunities for developing and sharing information essential for achieving sustainable development.**

All government agencies must improve the integration of federal, state, and local environmental data requirements, supporting information systems, and data-gathering capabilities. In particular, information systems should be built on an open architecture basis, allowing diverse users to extend capabilities by adding their own software or hardware interfaces and capabilities. Data residing in numerous locations can thus be assembled and processed by many users working with varied applications. Additionally, monitoring, processing, and delivery technologies must be better integrated. Efforts to calibrate, correlate, and validate measurement sources and application models should also continue.

Primarily because of potential cost savings, but also because of observational capabilities not before possible, there is significant regional and local government and commercial interest in data obtained through satellites and other remote sensing systems. Enhanced monitoring programs will aid the U.S. commercial environmental information industry in broadening its markets internationally.

Federal, state, and local governments currently have programs for gathering and sharing environmental data. For example, at the federal level, the Mission to Planet Earth program and the Global Earth Observing System Data and Information System of the National Aeronautics Space Administration provide data about the earth's land surface, water, ecosystem processes, oceans, and atmospheric chemistry to a broad range of users. The Interagency Federal Geographic Data Committee has established the National Spatial Data Infrastructure and associated clearinghouse to provide a means to inventory, document, and share geographic and spatial data. The National Environmental Data Index currently in development by The Department of Commerce's National Oceanic and Atmospheric Administration (NOAA) will provide data collection, integration, search, and index standards for environmental data.

The U.S. Department of Agriculture maintains a variety of ground-based environmental-monitoring networks for water quality and quantity, forest cover and condition, and other parameters. These systems provide raw data and "ground truth" for validating and enhancing satellite data.

The NOAA-supported Global Ocean Observing System and related CoastWatch program collect significant data of interest to those working on coastal and wetlands issues. The Department of the Interior's National Biological Service gathers, analyzes, and disseminates biological information needed for good stewardship of the nation's natural resources.

INITIATIVE **Through the INSIGHT 2000 initiative, the federal government will establish or broaden a series of partnerships to improve access to federally sponsored environmental data.**

The INSIGHT 2000 initiative is designed to bring together state, regional, and local governments, academia, industry, and decisionmakers to address environmental monitoring and information needs. The primary focus will be on making environmental data more accessible and useful to decisionmakers. In many cases the federal government will be the primary source of such data, while industry develops the software and other tools for providing data to decisionmakers.

INSIGHT 2000 will promote an integrated and comprehensive approach to the issues and opportunities associated with long-term sustainability at the local and regional levels. This initiative will build on current federal and local efforts such as the Global Change Data and Information System, National Spatial Data Infrastructure, National Environmental Data Index, and National Biological Information Infrastructure. Federal participation in the initiative will be coordinated through the Federal Geographic Data Committee and the National Science and Technology Council.

Sharing Information for Better Decisionmaking

The California Resources Agency has established the California Environmental Resources Evaluation System, providing a central coordinated location for sharing information about the state's natural resources. The system provides public access to environmental information about the state, including satellite imagery, census databases, environmental impact reports, and vegetation and wildlife databases. It also provides analytical tools and services such as searching capabilities, electronic data catalogues, and geographical information systems (GIS).

The city of Scottsdale, Arizona, is using GIS to aquire and integrate localized, short-term data with global, long-term data. Besides greatly enhancing the capability for local decisionmaking, the use of advanced GIS techniques, monitoring technology, and remotely sensed data are expected to save the city approximately $5 million in annual operating costs.

INITIATIVE **Systems will be established for disseminating and applying environmental information to enable effective environmentally related decisionmaking by businesses, regulators, and the public.**

Many technologies and techniques are commercially available today that are efficient, allow recycling of materials, or otherwise reduce or control waste and pollution. Smaller manufacturers in particular face difficult challenges in identifying and adopting technologies and techniques that can enable them to be environmentally sound and competitive. These challenges are associated with such factors as small staff size, intense pressure to ship products, lack of staff expertise on a given technical area, and lack of private consultants who serve smaller manufacturers. These factors often result in companies simply missing major opportunities.

> *"Widespread diffusion of existing off-the-shelf technologies could go a long way in further reducing pollution. However, many in industry, particularly small businesses, are unaware of pollution prevention options."*
>
> *— Office of Technology Assessment*

Environmental technology developers, companies that make use of environmental technologies, the financial community, regulators, and the public need timely, comprehensive, and carefully targeted information from sources they trust. Because a wide variety of organizations fill one or more roles in the generation and diffusion of technology-related information, increased cooperation is critical to the effectiveness of public and private efforts to disseminate and apply this information. This applies equally to information generated by government agencies and nongovernmental organizations.

Supporting Environmental Improvements in Small Businesses

In 1980, the Small Business Administration began developing a network of Small Business Development Centers across the United States. Today, more than 900 centers provide an invaluable source of information to small businesses through a network of 55 state, regional, and territorial programs.

One center, located in Nevada, has recently begun a Business Environmental Program to provide free and confidential assistance in hazardous waste management and multimedia pollution prevention. In 1994 alone, the program responded to more than 1,600 requests for information and provided training seminars to 1,000 individuals. A project is now underway to assess the capabilities of the Small Business Development Centers to deliver environmental assistance at a national level.

Part IV

A Call to Action

Chapter 4

A Call to Action

Moving forward effectively will mean moving forward together. Achieving our vision of sustainability will require that individuals and organizations recognize their unique ability to contribute to the goals set forth in this national environmental technology strategy. Industry, labor, communities, nongovernmental organizations, individuals, the states and federal government, and nations around the world all have important responsibilities in this effort to advance environmental technologies. The key to progress is building on the strengths of each sector in order to achieve goals collectively that cannot be achieved individually.

The agencies of the federal government are developing specific action plans for implementing the national environmental technology strategy embodied in this document. We urge others to join in this process. We invite industry, communities, the states, and nongovernmental organizations to develop their own action plans to advance environmental technologies.

The findings, goals, and initiatives outlined in this document can serve as an initial model. The appendix to this document contains a list of current federal programs that can provide additional guidance or aid in identifying ways to become involved. The action plans under development by the federal agencies can also serve as examples. In addition, the Interagency Environmental Technologies Office can help interested individuals and groups keep abreast of our nation's evolving environmental technology strategy.

Our challenge to you is this: Join us in building a more sustainable society through environmental technology. On Earth Day of 1996 let us share information and evaluate our progress.

The Role of Nations

As we move into the post-Cold War era, the United States is reorienting its foreign policy and economy to emphasize international objectives that have become compelling because of the changed world order. Through the strategy contained in this document, we seek to accelerate the development of a new generation of environmental technologies with far-reaching benefits, both domestically and globally. This process has already begun. Policies and international conventions to protect the planet, such as the Montreal Protocol, Vienna Convention, and Agenda 21, have generated demands for new environmental services and technologies. A new Climate Technology Initiative to stimulate development of new "climate-friendly" technologies was presented by the United States at the recent, March 1995, climate change conference in Berlin.

"We must be concerned about countries that are falling further and further behind. . . . Successful assimilation of the latest technologies . . . can allow countries to leapfrog certain investments and stages of development, leading to less pollution, higher productivity, and higher incomes."

— *Timothy E. Wirth*
Undersecretary of State for Global Affairs
U.S. Department of State

Demand for environmental technologies will increasingly be found outside the United States in countries with growing populations, rapid industrialization, and rising incomes. Thus, the international dimension of the national environmental tech-

CLIMATE TECHNOLOGY INITIATIVE

The Administration's Climate Technology Initiative is a linked set of domestic and international measures to accelerate the development and diffusion of climate-friendly technologies. Any solution to the climate change problem will rely heavily on technologies. Barriers to the use of these technologies include lack of information, economies of scale, problems with financing, and the need for further innovation and technological development.

The initiative addresses these barriers with a variety of activities such as the transfer of voluntary programs to other governments and nongovernmental entities, a pilot project to evaluate innovative financial mechanisms, an implementation study for a network of regional sustainable energy centers, financial and technical support for the development of national climate change action plans, and regional initiatives to help developing countries and economies in transition gain control of greenhouse gas emissions.

nology strategy must encourage U.S. partnerships with these nations to adopt sustainable development practices and employ appropriate technologies to meet these challenges. These partnerships should embrace a range of activities including information sharing, capacity-building for better assessment and use of modern technologies.

It is clear that we are not alone in this endeavor. In fact, if we are to succeed in creating a more sustainable planet, it can only be with the help and cooperation of other nations. We invite other nations, and communities of nations, to join us in advancing environmentally critical technologies and practices worldwide. An international conference on environmental technologies would help as a first step in coordinating strategies and programs globally.

The Federal Role

The ultimate responsibility for achieving the vision embodied in this national environmental technology strategy by Earth Day 2020 lies with individual citizens, businesses, and communities across America. However, the federal government serves an important catalytic and coordinating role. In some areas the government will play a leading part, helping to guide industry and the states. In other areas, it will be a partner facilitating the leadership role taken by others. The Congress will be crucial in authorizing programs and appropriating funds to support federal and state activities.

Under this national environmental technology strategy, federal, state, and local governments will work together closely and take the lead in the areas of regulatory innovation. This role will be accomplished in partnership with industry, the financial community, nongovernmental organizations, and the public. The federal government will also support research and development, along with industry and academia, which is key to developing future generations of environmentally benign products, processes, and services. Finally, the government will actively adopt sound environmental management practices within its own operations and facilities and procure more environmentally benign products.

The federal government will aid industry, along with the states, financial community, and nongovernmental organizations, in the area of promoting the development of overseas markets. Industry will take the leading role in this area, and the federal government will be a supporting partner.

The Role of Industry

We challenge the industries of America to serve as the primary drivers in implementing this strategy. It is industry that owns, develops, implements, markets, and uses the environmental technologies that protect the environment and help fuel economic growth. We also call upon industry to take the lead, along with nongovernmental organizations, in building a more sustainable society through environmental technology.

In this context, industries include not only traditional manufacturing companies, but also firms that provide services and produce our food, housing, and energy. Many of these businesses are already incorporating techniques for pollution avoid-

ance and waste reduction into their production systems. These activities increase productivity and save money.

Widespread recognition within all sectors of our economy that environmental protection is good business is crucial to attaining sustainable development. Over the next 25 years, industries will make *all* technology cleaner, not just clean technologies.

Industry has an important role to play in developing standards and best practices, which can be instrumental in facilitating the widespread use of new technologies and processes. For example, the American National Standards Institute and the Canadian Standards Organization are currently producing standards related to environmental management systems, auditing, and performance evaluation.

Entrepreneurs in industry will develop and employ the innovative environmental technologies needed to move from our nation's command-and-control regulatory system to an incentive-based system with high standards and maximum flexibility. These new technologies will help open new overseas markets and provide competitive advantages for U.S. industries.

The Role of Communities

Cities, towns, counties, and community organizations have learned in recent years that tremendous economic and environmental benefits are associated with applying technologies appropriately. Technologies that advance community sustainability are driven by shared perceptions of problems and a search for economically viable and equitable solutions. The largest and most complex class of environmental technologies are those supporting our communities: technologies that transport people or goods, produce and deliver energy, treat water supplies and waste products, provide food, and route and process information. To achieve sustainability, technological solutions must be integrated with the unique economic, social, political, and cultural circumstances of each community.

> *"Technology is an essential part of making our nation's communities more sustainable."*
>
> *— J. Gary Lawrence*
> *Director*
> *Center for Sustainable Communities*
> *University of Washington*

Community sustainability is the end result that we must seek to deliver through technology. Basic to this delivery process — one already well underway in the United States — will be an expanded grasp of the concept of "infrastructure." This term is traditionally applied to urban, suburban, and rural physical networks (utilities, water supply, roads, and rails). Expanded to "sustainable infrastructures," the concept encompasses all physical systems that support modern civilization, whether built environment or natural systems. To be sustainable, these infrastructures must be distributed equitably, supporting the economic activities and well-being of all people in the community, whether present populations or future generations.

We call upon communities throughout America to join together in laying the groundwork so that more sustainable activities and infrastructures will be in place in our cities, towns, and rural areas by Earth Day 2020.

The Role of the States

Many state governments are now actively working to promote the development and application of environmental technologies. The strategy contained in this document is built on lessons provided by these "laboratories of democracy" and is designed to encourage and complement these activities. In 1994, for example, California developed a strategic plan for promoting its environmental technology industry. The California strategy focuses on preserving and promoting the state's high environmental standards, encouraging a shift toward avoidance technologies, and assisting California-based companies in advancing environmental systems, goods, and services. The California strategy emphasizes partnerships among government, industry, academia, nongovernmental organizations, and interest groups.

Similarly, in North Carolina the state government, universities, and industry associations are joining together to establish an Environmental Technologies Consortium. This nonprofit organization will facilitate the development, manufacture, use, and export of environmental technologies. The consortium will support research; coordinate identification of manufacturers willing to provide demonstration or validation sites for new technologies; work to develop a state regulatory climate conducive to innovation; and promote regional networks among investors, entrepreneurs, and potential users to speed the development of new environmental technologies.

State actions are crucial because primary responsibility for the implementation of programs mandated by federal law lies with the states.

At the multistate level, the Western Governors Association is identifying barriers and developing approaches to speed the commercialization of environmental technologies by improving state permitting programs and identifying sites to demonstrate and evaluate technologies. The National Governors Association is working closely with the western governors to expand the scope of state programs.

The Role of the Financial Community

A shortage of capital in the demonstration and scale-up stages is a major impediment to the widespread use of innovative environmental technologies. Because of the uncertainty associated with environmental markets, those with venture capital often regard environmental technologies as poor investments. The financial community has an important role to play in working with industry, government regulatory agencies, and community groups in helping reform the regulatory system and remove the many barriers that can make the markets for environmental technology unfriendly to innovation and investment

Increasingly, private sector capital providers are part of a broader process to develop reforms that will increase the attractiveness of environmental technologies as an investment. We urge the financial community to continue this valuable interaction and develop new and innovative strategies to link good ideas with the capital necessary to bring them to market.

"In order to make progress, we need to engage the entrepreneurial energy, inventiveness, and creativity of the industrial sector and the interest of the finance sector. One key element of the strategy has to be to create the reliable expectation that success will be rewarded with profits, that it will be profitable to be green, and it will be even more profitable to be greener in smarter ways."

Jonathan Lash, Co-Chair, President's Council on Sustainable Development December 13, 1995

The Role of Nongovernmental Organizations

Nongovernmental organizations are key partners in achieving this vision for the future. Trade and professional associations, environmental groups, community groups, and international organizations can help link and enhance the activities of industry, government, academia, and individual citizens. Nongovernmental organizations can build trust and bridge differences between industry, policymakers, and the community. At the local level, many environmental organizations work with communities, neighborhood groups, and local officials to develop more sustainable communities.

Nongovernmental organizations can facilitate the partnerships needed to develop a new approach for environmental protection. The Environmental Defense Fund, for instance, created a task force on solid waste reduction with McDonald's

Corporation. The task force developed a comprehensive plan for solid waste reduction — including source reduction, re-use, recycling, and composting — that could reduce the company's waste stream by 80 percent. In 1991 McDonald's began implementing this plan throughout the company's thousands of restaurants, distribution centers, and suppliers.

Nongovernmental organizations can help educate and motivate members. The Air and Waste Management Association provides information to its membership about sustainable development and design-for-environment techniques. Organizations such as the American Association of Engineering Societies and the Civil Engineering Research Foundation will continue to play an important role in achieving a consensus within the engineering community on approaches to attaining sustainable development. Environmental groups such as the Sierra Club, Natural Resources Defense Council, and National Wildlife Federation, and think tanks such as the World Resources Institute, educate people about the value of protecting the environment and motivate environmentally conscious behavior. These and other community groups also play an oversight role, ensuring that industry and government address environmental concerns that are of particular concern to the public.

Nongovernmental organizations also play an important role in other areas such as research and development. For example, The Nature Conservancy and the U.S. Department of the Interior's National Biological Service are working together to provide better information on the status of biological resources by integrating data developed at the state level, and improving access to that information by all parties. The American Forest and Paper Association and the U.S. Department of Energy have established a partnership for research and development based on the paper industry's 2020 Vision, which includes sustainability concepts.

The challenge for nongovernmental organizations is to develop innovative approaches, build trust, and inform decisions in both the public and private sectors.

The Role of Teachers and Academia

Our attitudes towards technologies and the environment are shaped through our educational system. Teachers and scientists in schools, state universities and colleges, community colleges, and research institutions play a key role in advancing environmental technologies. Individuals in these institutions educate our young people, perform necessary environmental research, and train the nation's workforce. Research institutions develop new scientific and economic tools, collect and analyze data necessary for understanding material and energy flows, and develop new techniques for life-cycle assessments and product design.

"Our communities must construct a system which embraces economic development along with resource conservation and environmental quality — sustainable development is the framework for such a system."

— Linda Giannelli Pratt
County of San Diego, Department of Environmental Health, Pollution Prevention Program

Universities and colleges can restructure their curricula to offer additional cross-cutting, multidisciplinary courses for students majoring in engineering, business, and the social sciences. For example, the National Pollution Prevention Center for Higher Education, located at the University of Michigan, is forging collaborative efforts among industry, government, nongovernmental organizations, and academia to educate citizens on the concepts and techniques of pollution prevention. The objective is to provide information and educational materials that integrate these concepts into traditional disciplinary studies such as accounting, chemical engineering, industrial design, and business law.

Community colleges also play a central role in environmental education and training. Because these colleges regularly interact with local industries in developing curricula to support training needs, they offer an excellent network for exchanging information on sustainable development.

The challenge to academic institutions is to expand environmental education and research, multidisciplinary courses on pollution prevention, and curricula for training the nation's workforce. In general, all teaching institutions have a critical role to play in increasing public understanding of environmental and technological issues in our society.

The Role of Employees and Organized Labor

As employees and workers, many of us are directly responsible for the operation of technologies and have opportunities to influence process development, maintenance, and management strategies in the workplace. Significant variations in the environmental and energy performance of firms are often related to operator practices and can be improved if workers are fully integrated into the environmental management process. Employees are often in the best position to identify ways to save energy and materials while increasing product value.

Environmental technologies hold tremendous market and employment opportunities, including new jobs for workers in industry and opportunities to improve environmental and economic performance, workplace health and safety, and community environmental standards. An example of a labor-management environmental partnership is the Waste Elimination and Cost Awareness Reward Everyone (WE CARE) program of the United Automobile Workers and General Motors Corporation. By working together, setting goals, tracking progress on a regular basis, and communicating results, employees in this program enhance the conservation of resources and the reduction of pollution from their facilities. Organized labor can help facilitate this important role of workers and aid employees in benefiting from the productivity gains they help produce.

The Role of the Individual American

Every citizen has a crucial role to play in achieving the vision embodied in this strategy. As members of communities, we can work together to establish environmental and economic goals and develop strategies to attain them. As consumers, we can dramatically alter U.S. consumption of resources through our product choices and demands for better product information. Every American can make a difference by conserving resources; buying energy-efficient vehicles, appliances, and homes; and becoming involved in efforts to encourage the development of sustainable communities. Such simple everyday acts are in keeping with the spirit of the first Earth Day. By continuing and enhancing this spirit, we can ensure lasting economic growth while protecting the environment.

The Road to Earth Day 2020

Thousands of individuals have contributed to this strategy. We have heard the views of representatives of small and large companies and of labor. We have listened to officials from state and local governments. We have met with individuals from environmental groups, universities and community colleges, banks and investment firms, and we have consulted with federal agencies. The strategy embodied in this document is the outcome of a truly participatory effort. The strategy is stronger because it emanates from that diversity of viewpoints.

Movement in the direction of a more sustainable society by Earth Day 2020 will benefit all of us. The challenges facing us in the next 25 years from population and economic growth are unprecedented in human history. Our capacity to meet these challenges will require new approaches to working together. Trust is the key to success. Whether our vision for Earth Day 2020 becomes a reality is up to each of us.

In the years ahead, it will be important to re-visit this vision and the strategy for achieving it. Both will evolve as our knowledge and technologies mature, as environmental problems change, and, especially, as we ourselves readdress the values by which we live. The strategy will need to be amended as we move forward.

We will also need to be far-sighted. If our environmental technology industry is to remain competitive in the global marketplace, we must implement actions today that will be responsive to tomorrow's problems, needs, and opportunities. To meet future challenges will require policies that have the capacity and flexibility to move us significantly beyond where we are today. We will have to take advantage of markets in a highly mobile way. That means moving information, people, and capital rapidly. Most of all, our nation's human resources must be developed continuously and applied effectively. To achieve that, we as individuals must achieve a more sophisticated understanding of our environment and economy.

The question is where do we go next. In the end, that is up to each of us. All Americans have an important role to play in building a bridge to a sustainable future.

APPENDICES

APPENDIX 1

Selected Federal Programs Related to Environmental Technologies

Program Name	Brief Description	Point of Contacts for More Information
Interagency Programs		
Interagency Environmental Technologies Office (IETO)	This interagency office is a focal point for private sector information exchange; interagency technology collaboration; and regulatory, legal, and procurement barrier reduction.	Interagency Environmental Technologies Office (202) 426-2078 (202) 426-2303 fax
Rapid Commercialization Initiative (RCI)	An interagency program to reduce commercialization barriers for environmental technologies by identifying demonstration sites, verifying results, and expediting permitting.	Department of Commerce (202) 482-2389 (202) 482-6274 fax IETO (202) 426-2079 (202) 426-2303 fax
INSIGHT 2000	An interagency federal government initiative to improve access to and use of federally sponsored data and information systems. The project includes partnerships and demonstrations with state and local governments and the private sector.	Committee for Environment and Natural Resources (CENR) Task Force on Observations and Data Management (202) 358-0258 (202) 358-3098 fax Internet: atuyahov@mtpe.hq.nasa.gov
Joint Subcommittee on Environmental Technologies (JSET)	This subcommittee of the National Science and Technology Council works to increase coordination and alignment of federal environmental technologies R&D and demonstration of environmental technologies.	National Science and Technology Council Joint Subcommittee on Environmental Technologies (202) 426-2078 (202) 426-2303 fax jwaddell@nsf.gov
Export Market Plans	The Trade Promotion Coordinating Committee's Environmental Trade Working Group (ETWG) is implementing a government-wide strategic plan for federal environmental technology export promotion. The ETWG has identified high-growth markets for U.S. environmental technologies exports. Plans are available on the National Trade Data Bank (NTDB) or through the National Technical Information Service (NTIS).	Department of Commerce International Trade Administration Office of Environmental Technology Exports (202) 482-5225 (202) 482-5665 fax oete1@ita.doc.gov
Global Change Research Program	An interagency program for national and international long-range global change observation and assessment. A major component is the Global Change Data Information System (GCDIS).	Global Change Research Office (202) 651-8250 (202) 554-6715 fax
National Environmental Data Index (NEDI)	An interagency effort to provide the framework and standards for data access, collection, integration, search, and indexing for federal environmental databases.	National Oceanic and Atmospheric Administration Environmental Information Services (202) 606-4090 (202) 606-0509 fax jamerson@esdim.noaa.gov

Program Name	Brief Description	Point of Contacts for More Information
Department of Agriculture Programs		
Environmental Technologies for Ecosystem Management	A U.S. Forest Service program that considers the application of environmental technologies in ecosystem management.	U.S. Forest Service (202) 205-1702 (Sustainable Forestry) (202) 205-1530 fax (202) 205-1793 Ecosystem Management) (202) 205-1012 fax
New Uses, New Products	The center supports several programs that research, develop, and demonstrate new sources of income for the agricultural community.	Alternative Agricultural Research and Commercialization Center (202) 690-1633 (202) 690-1655 fax
Energy Efficiency	The service develops hybrid electrical generating systems using renewable energy sources such as wind, solar, and biofuels.	Agricultural Research Service (806) 356-5734 (Renewable Energy) (806) 356-5750 fax (309) 681-6541 (Biofuels) (309) 681-6682 fax
Water Quality Program	The service provides education and technical assistance to farmers, ranchers, and foresters; R&D on improved agricultural and forestry management systems; as well as database development on agricultural chemical use and related farm practices.	Agricultural Research Service (301) 504-6441 (301) 504-5467 fax
Agricultural Monitoring Networks	USDA-maintained environmental monitoring networks for snowpacks, water quality/quantity, forest coverage/condition and other parameters..	Agricultural Research Service (301) 504-6822 (301) 504-5041 fax
U.S. Agency for International Development Programs		
Center for Environment	Providing support to more than 40 field missions in different countries, the Center works with public, private, and nongovernmental organizations to make investments in energy, natural resources, and urban infrastructures. The Center has initiatives to promote sustainable cities, climate change adaptation and mitigation, environmental technology, and conservation of biodiversity.	USAID, G/ENV (703) 875-4205 (703) 875-4639 fax
Center for Trade and Investment Services (CTIS)	The center provides the U.S. business and development communities with information on business opportunities in USAID-assisted countries. This includes links with entrepreneurs overseas to facilitate joint ventures and partnerships, and information on USAID-financed procurement opportunities.	USAID, G/EG/CTIS (800) 872-4348 (202) 663-2660 (202) 663-2670 fax ctis@usaid.gov
U.S.-Asia Environmental Partnership (USAEP)	The partnership works in 34 Asia/Pacific nations to match environmental needs with U.S. environmental experience, technologies, and practices. It involves matching grants, identification of project opportunities, dissemination of information on market and environmental needs, business exchanges, training, and professional networks.	U.S.-AEP Program/TR&D (800) 872-8723 (202) 482-0543 (202) 835-0333 (202) 835-0366 fax
Department of Commerce Programs		
Manufacturing Extension Partnership	The partnership is a national network of not-for-profit technical assistance organizations and national information resources that works to increase the competitiveness of U.S. manufacturers by enabling them to adopt improved technologies and techniques rapidly.	Manufacturing Extension Partnership Office (301) 975-3591 (301) 963-6556 fax http://www.nist.gov

Program Name	Brief Description	Point of Contacts for More Information
Department of Commerce Programs		
Environmental Technology Exports	The office facilitates increased exports of U.S. environmental technologies through cooperative initiatives with the private sector including market analysis, business counseling, industry outreach, trade promotion, and advocacy.	International Trade Administration Office of Environmental Technology Exports (202) 482-5225 (202) 482-5665 fax oete1@ita.doc.gov
Environmental Data Management System	The system provides environmental data and information about the atmosphere, ocean, earth, and sun to meet the needs of decisionmakers, scientists, and other users through the National Climatic Data Center, National Oceanographic Data Center, National Geophysical Data Center, and other centers of data.	National Oceanic and Atmospheric Administration (301) 457-5214 (301) 736-5828 fax NOAA Homepage: http://www.noaa.gov
Global Ocean Observing Systems (GOOS)	GOOS is an internationally coordinated system for operational data collection, analysis, and exchange of ocean observations and applications information. It is led by the National Oceanic and Atmospheric Administration.	GOOS Interagency Program Office (301) 713-2981 (301) 713-4392 fax usgoos@noaa.gov
Coastal Forecast System	The system improves our ability to measure, understand, and predict coastal environmental phenomena that impact public safety and well-being, the national economy, and environmental management.	National Oceanic and Atmospheric Administration Coastal Ocean Office (301) 713-3338 (301) 713-4044 fax
Coastwatch Program	The program delivers NOAA satellite environmental data and products for near real-time monitoring of U.S. coastal waters to support resource management and associated research.	National Oceanic and Atmospheric Administration Office of Satellite Data Processing and Distribution (301) 457-5120 (301) 457-5184 fax
Department of Defense Programs		
Environmental Security Technology Certification Program (ESTCP)	The program demonstrates and validates the most promising innovative technologies that meet urgent DOD environmental needs and environmental security objectives.	DOD Environmental Technology Office (703) 756-0790 (703) 756-1176 fax
National Environmental Technology Demonstration Program — National Test Sites	The program aids DOD and DOE environmental cleanup by providing test platforms and infrastructure, characterizing sites, and supporting public relation development.	DOD program office (703) 696-2117 (703) 696-2114 fax
Strategic Environmental Research and Development Program (SERDP)	The joint DOD, DOE, and EPA program identifies and develops technology to meet DOD environmental needs. Participants include federal agencies, academia, and private industry.	SERDP Program Office (703) 696-2117 (703) 696-2114 fax
Department of Energy Programs		
Energy Efficiency	A wide range of programs to increase the efficiency of energy use. Examples include technology development for transportation, buildings, and industry.	Office for Energy Efficiency and Renewable Energy (800) DOE-EREC http://www.eren.doe.gov
Environmental Management	The office is responsible for innovative R&D on plumes contamination, high-level waste tanks, land fills, mixed waste, and decontamination and decommissioning for cleaning up weapons manufacturing sites.	Office of Technology Development (800) 736-3282

Program Name	Brief Description	Point of Contacts for More Information
Department of Energy Programs		
Fossil Energy	The office oversees a national partnership program to develop technologies that can ensure environmentally clean, affordable energy from coal, petroleum, and natural gas.	Office of Fossil Energy (202) 586-1650 (Coal Technologies) (202) 586-1369 (Natural Gas and Petroleum Technologies) (202) 586-1577 (Advanced Research and Special Technologies)
Climate Challenge Program	A voluntary program that helps utilities and industry find innovative ways to reduce emissions of greenhouse gases.	Office of Utility Technologies Office of Energy Efficiency and Renewable Energy (800) OOE-EREC http://www.eren.doe.gov
Renewable Energy	A wide range of programs to increase the cost-effectiveness of renewable energy sources including photovoltaic and solar thermal energy; biomass power; geothermal electric and heat pumps; wind energy; integrated resource planning; hydrogen production, storage, and use; and addresses increased efficiency of transmission and distribution incorporating superconductors; and energy storage.	Office for Energy Efficiency and Renewable Energy (800) DOE-EREC http://www.eren.doe.gov
Environmental Protection Agency Programs		
Project XL	A program through which the federal government will cooperate with the states and selected firms to test on a pilot basis a new regulatory approach that will provide incentives for industry to find innovative ways to achieve environmental goals.	Office of Policy, Planning, and Evaluation (202) 260-4034 (2020 260-0780 fax
Environmental Technology Initiative (ETI)	Forms multi-organizational partnerships to stimulate development and use of innovative technologies.	ETI Hotline (202) 260-2686 (202) 260-2685 fax
Common Sense Initiative (CSI)	Works with selected industries and stakeholders to improve environmental results through regulatory flexibility and the use of pollution prevention technologies.	Office of Air and Radiation (202) 260-8953 (202) 260-9766 fax
Design for the Environment (DFE)	Encourages businesses to incorporate environmental considerations into the design of products and manufacturing processes.	Pollution Prevention Information Clearinghouse (202) 260-1023 (202) 260-0178 fax
33/50	Encourages voluntary reduction in releases of 17 toxic pollutants from industrial sites.	Office of Prevention, Pesticides, and Toxic Substances (202) 260-6907 (202) 260-2219 fax
Green Lights and Energy Star Buildings	Offers technical assistance and recognition for producing and applying energy reduction measures in buildings, computers, and lighting.	Green Lights and Energy Star Hotline (202) 775-6650 (703) 934-3183 fax
Superfund Innovative Technology Evaluation (SITE)	Encourages commercialization of innovative technologies for treating hazardous waste contamination through demonstration, emerging technology, and monitoring and measurement programs and technology transfer.	National Risk Management Research Laboratory (513) 569-7861 (513) 569-7620 fax

Program Name	Brief Description	Point of Contacts for More Information
Environmental Protection Agency Programs		
GW Green University Initiative	A public-private partnership between EPA and George Washington University to develop the nation's first model "green" university and to collaborate with interested organizations worldwide.	Office of Administration and Resources Management (202) 260-1647 (202) 260-0215 fax
U.S. Technology for International Environmental Solutions (U.S. TIES)	Uses technical assistance and training, information exchange, and technology demonstration to match environmental problems abroad with U.S. suppliers of technical solutions.	Office of International Activities (202) 260-4894 (202) 260-4470 fax
Department of Health and Human Services		
NIEHS Superfund Basic Research Program	Basic and applied R&D on technologies and strategies to reduce and monitor exposure to toxic substances, including bioremediation, steam injection to remove solvents from soil, combustion and super critical wet oxidation.	National Institute of Environmental Health Sciences (301) 496-2919 (301) 496-0563 fax
Monitoring and Assessment of Human Exposure	Basic and applied molecular biology to develop biomarkers of exposure, effect, and susceptibility.	National Institute of Environmental Health Sciences (301) 496-2919 (301) 496-0563 fax
Department of Housing and Urban Development Programs		
Building Systems Technology	Basic R&D on innovative and good practices for a full range of housing types.	Office of Policy Development and Research (202) 708-4370 (202) 619-8360 fax
Empowerment Zones and Enterprise Communities	A program to empower local citizens to become more involved in their communities and aware of environmental technology benefits that result from attracting attract new business and industries to depressed areas.	E-2 and E-C Program (202) 708-2035 (202) 708-7543 fax Department of Agriculture Empowerment Zone and Enterprise Community Program (202) 619-7982 (202) 401-7420 fax
Land Use Systems Technology	Research and technical aspects of urban development, land use, open space, environmental protection, and other contexts of development.	Office of Research, Evaluation, and Monitoring (202) 708-0574 (202) 619-8360 fax
Sustainable Communities Development Systems	Overall policy and technical purview of technologies affecting all dimensions of the ecological, land, natural resources, industrial, and development aspects of urbanization.	Office of Community Planning and Development (202) 708-1911 (202) 708-3363 fax
Department of the Interior Programs		
National Spatial Data Infrastructure (NSDI) and National Geospatial Data Clearinghouse (NGDC)	Funding for cooperative agreements to state/local government agencies and private organizations. Electronic network with information about geospatial data.	U.S. Geological Survey (703) 648-5514 gdc@usgs.gov
National Biological Information Infrastructure	Develops and promotes measures that make data and information on biological resources, biological expertise, and analytical tools more accessible and useful.	National Biological Service NBII Coordinator (202) 482-3980 (202) 273-3279 fax
Appalachian Clean Streams Initiative	Facilitates and coordinates federal, state, and local government agencies involved in cleaning up streams polluted by acid drainage, primarily from abandoned coal mines. Program builds on local and watershed initiatives.	Office of Surface Mining (202) 208-2782 (202) 219-3106 fax

Program Name	Brief Description	Point of Contacts for More Information
Department of the Interior Programs		
National Earthquake Hazards Reduction	Provides real time data on earthquakes and impacts to buildings and highway structures. Data is used by emergency preparedness managers and for developing construction standards and land use planning.	U.S. Geological Survey (303) 273-8500 (303) 273-8444 fax
National Water Quality Assessment Program	Conducts an ongoing assessment of surface-water and ground-water quality in 60 large watershed or aquifer systems nationwide. Program goals are to describe the status and trends to natural and human factors.	Department of the Interior U.S. Geological Survey (703) 648-5012 http://wwwrvares.er.usgs.gov/nawqa/nawaq_home.html
Department of Labor Programs		
Skills and Training for Environmental Technology	A joint DOL and DOE program to empower employees to reduce business costs through improved energy efficiency and waste reduction. Includes job skills training, dissemination of skills information, and model training curricula.	Employment and Training Administration Office of Policy and Research (202) 219-8660 (202) 219-5455 fax
National Aeronautics and Space Administration Programs		
Earth Observing Commercial Applications (EOCAP) and Visiting Investigator Program (VIP)	EOCAP enables commercial developers of advanced remote-sensing equipment and techniques to test and validate their concepts using government facilities and test beds. VIP provides additional partnership opportunities with industry to develop commercial remote-sensing/GIS demonstration projects.	Stennis Space Center (601) 688-1814 (601) 688-7455 fax
Mission To Planet Earth	Earth observing data: space, land surface, coastal, ecosystem processes, ice, earth, oceans, and atmospheric information, including wide accessibility through the Earth Observing System Data and Information System (EOSDIS) and Distributed Active Archive Centers (DAAC).	Goddard Space Flight Center (301) 286-5520 Internet: john.dalton@gsfc.nasa.gov http://spso.gsfc.nasa.gov/spso_homepage.html
The Pathfinder Program	Makes research-quality global change data sets more easily available to the science and environmental communities for use in research and assessments.	Earth Science Data and Information System Branch (202) 358-0759 (202) 358-3098 fax mmaiden@mtpe.hq.nasa.gov
Remote Sensing Public Access Center	Focuses on developing pilot end-user remote-sensing database applications and applying new digital library technologies that can enable and demonstrate the application and accessibility of earth science databases.	Remote Sensing Public Access Center (304) 387-8239 (304) 367-8211 http://rspac.lvv.nasa.gov
National Science Foundation Programs		
Environmentally Benign Chemical Synthesis and Processing	Encourages research on new approaches that will support the clean, efficient, and profitable industries of tomorrow in the production of plastics, pharmaceuticals, and electronics.	National Science Foundation (703) 306-1840
Environmentally Conscious Manufacturing	Supports research that addresses the environmental impact of decisions made at various stages of a product's life from conception to design, raw materials consumption, synthesis, processing, use, recycling, and/or disposal.	National Science Foundation (703) 306-1330
Technology for a Sustainable Environment	A joint program with EPA to advance the development and use of innovative technologies directed at avoiding environmental harm.	National Science Foundation (703) 306-1370

Program Name	Brief Description	Point of Contacts for More Information
Department of State Programs		
Department of State Foreign Affairs Network (DOSFAN)	In cooperation with the University of Illinois at Chicago, an Internet-based "electronic reading room" for Department of State information dealing with issues related to international environmental affairs, population, and sustainable development.	Office of Public Affairs (202) 647-6681 http://dosfan.lib.uic.edu.dosfan.html
U.S. Man and the Biosphere Program (U.S. MAB)	Complementing the International MAB Program founded through the United Nations in 1970, the U.S. MAB supports policy-relevant research by economists, biologists, geographers, and other scientists addressing the interaction of communities and complex systems.	Department of State Executive Secretariat for the U.S. Man and the Biosphere Program (202) 466-1935
International Coral Reef Initiative (ICRI)	Multi-agency and international program to promote the sustainable use of coral reefs and related ecosystems through improved scientific understanding, improved management techniques, and international capacity-building.	Department of State Office of Ecology and Terrestrial Conservation Bureau of Oceans and International Environmental and Scientific Affairs (202) 647-0658 sdrake@state.gov
Interagency Programs (continued)		
INSIGHT 2000	An interagency federal government initiative to improve access to and use of federally sponsored data and information systems. The project includes partnerships and demonstrations with state and local governments and the private sector.	Dept. of Interior/USGS) Federal Geographic Data Committee (703) 648-5514 (703) 648-5755 fax Internet: gdc@usgs.gov Homepage: http://fgdc.er.usgs.gov).
National Spatial Data Infrastructure	An interagency effort that develops and promotes policies, standards, partnerships, and technology to improve access and use of geographic and spatial data.	Dept. of Interior/USGS) Federal Geographic Data Committee (703) 648-5514 (703) 648-5755 fax Internet: gdc@usgs.gov Homepage: http://fgdc.er.usgs.gov).
Environmental Protection Agency Programs (continued)		
Environmental Monitoring and Assessment Program (EMAP)	A nationwide initiative to assess and document the status and trends in the condition of the nation's ecological resources.	Office of Research and Development (202) 260-5776 (202) 260-0929 fax http://earth1.epa.gov/emap/
Internet/Public Access Initiative	Provides EPA data and information for public access over Internet.	Office of Information Resources Management (202) 260-4465 (202) 260-0835 fax http://earth1.epa.gov/

Appendix 2

Technology for a Sustainable Future

Conferences, Forums, and Workshops

This strategy builds heavily on the deliberations from the following meetings, which were organized by the White House Office of Science and Technology Policy and the National Science and Technology Council, often in partnership with other organizations. In those cases where the federal government was not the primary organizer, the responsible organizations are noted in parentheses.

1994

July 15	Technology for a Sustainable Future "Kickoff" Event, Golden,CO
August 26	Forest Products Information Exchange, Atlanta, GA
August 30	Forum on Technology for a Sustainable Future, Seattle, WA
September 7-9	Second Annual Private Enterprise-Government Interactions (PEGI) Task Force Roundtable Conference, Herndon, VA
September 20	Overcoming Financial Barriers to the Development of Environmental Technologies, New York, NY
September 26	Promoting the Development and Use of Innovative Remediation Technologies, Pasco, WA
September 28-	Partnerships Supporting Education About the Environment, The Presidio of San October 1, Francisco, CA
October 4-5	Chemical and Petroleum Refining Industries, Houston, TX
October 6-7	Residential Construction, Pittsburgh, PA
October 11	Manufacturing for a Sustainable Future, Albuquerque, NM
October 17	Service Industries and the Environment, Washington, DC
October 20-21	Prospecting for Opportunities in Energy and Materials Flows: Policy Options to Promote Environmental Technologies, San Francisco, CA
October 24	Forum on Texas, Mexico, and Expanding International Markets for Environmental Technologies, El Paso, TX
October 28	Sustainable Technologies for the Americas, Miami, FL
October 31-	Sustainable Food Production: Policy Options to Promote November 1 Technologies, Kansas City, MO
November 9	Promoting Sustainable Urban Infrastructure, Los Angeles, CA
November 14	Federal Initiatives to Promote Environmental Technologies Development, Research Triangle Park, NC
November 15-16	West Coast Remediation Marketplace: Business Opportunities for Innovative Technologies, San Francisco, CA

November 21	Commercializing Environmental Technologies, Washington, DC
November 29	Promoting Environmental Technologies in the Pulp and Paper Industry, Burlington, VT
December 11-13	White House Conference on Environmental Technology, Washington, DC

1995

January 17-18	Environmental Monitoring, San Diego, CA
February 1-2	Public-Private Partnership for Sustainable Use of Natural Resources, Boise, ID
February 2	Creating Agents of Change (American Association of Engineering Societies), Washington, DC
February 24	Forum on the Rapid Commercialization Initiative (Civil Engineering Research Foundation and Hazardous Waste Action Coalition), Washington, DC
March 8	Sustainable Cities, Philadelphia, PA
March 16	Green Engineering Roundtable (Virginia Polytechnic University), Falls Church, VA
March 24	Small Business Manufacturing for Environmental Technology, Kansas City, MO
March 28	Electric Power for a Sustainable Future, Morgantown, WV
March 29	Rapid Commercialization Initiative Industry Workshop, Gaithersburg, MD
April 6	Verification (National Governors' Association), Research Triangle Park, NC

For sale by the U.S. Government Printing Office
Superintendent of Documents, Mail Stop: SSOP, Washington, DC 20402-9328
ISBN 0-16-048012-4

APPENDIX 3

Key References Used in Generating Tables

Title, Year	Code
Annual Energy Outlook 1995	DOE/EIA-0383(95)
Annual Energy Review 1993	DOE/EIA-0384(93)
National Air Pollutant Emissions Estimates, 1900-1991	EPA/454/R-92-013(92)
National Air Quality and Emissions Trends Report, 1993	EPA/454/R-94-026(94)
Statistical Abstract of the United States 1994	Bureau of the Census
National Transportation Statistics, Historical Compendium, 1960-1992	DOT-VNTSC-BTS-93-1
Agricultural Resources and Environmental Indicators	USDA Ag Handbook, #705(95)
Statistical Abstracts of the United States 1994	USDA

Other References

As part of the development of this strategy, two studies were undertaken by the National Commission for Employment Policy to examine the job impacts of environmental technology policies and investments:

Environment and Jobs: The Employment Impacts of Federal Environmental Investments. Skip Laitner, Marshall Goldberg, Michael Skeehan, with Mark Baldwin.

Promoting Growth and Job Creation through a New Generation of Environmental Technologies. J. Andrew Hoener, Alan S. Miller, Frank Muller.

Copies are available through the:
National Commission for Employment Policy
1441 L Street, NW Suite 9000
Washington, DC 20005

Photography Credits

Cover: Golden Gate Bridge © 1994 Paul Grabhorn; Bhutanese Girl © 1993 Grabhorn; Windmills © 1994 Grabhorn; Globe, NASA. *Table of Contents*: © 1995 Grabhorn (1,4); Department of Energy (2); NASA (3). *Page 1*: © 1994 Grabhorn. *Page 2*: © 1994 Grabhorn. *Page 6*: NASA. *Page 7*: DOE. *Page 8* *EPA*. *Page 10*: DOE. *Page 19*: DOE. *Page 23*: EPA. *Page 24*: DOE. *Page 25*: NASA. *Page 26*: DOE. *Page 27*: NASA (1,4); © 1995 Grabhorn (2,3). *Page 28*: NASA. *Page 33*: DOE. *Page 37*: © 1993 Grabhorn. *Page 40*: NASA. *Page 42*: DOE. *Page 50*: © 1994 Grabhorn. *Page 51*: National Park Service. *Page 52*: © 1987 Grabhorn. *Page 54*: University of Maryland at College Park. *Page 65*: © 1994 Grabhorn. *Page 66*: © 1994 Grabhorn. *Page 68*: NASA. *Page 69*: DOE. *Page 72*: ©1994 Grabhorn. *Page 75*: © 1995 Grabhorn. *Page 77*: ©1993 Grabhorn. *Back Cover*: ©1992 Grabhorn.

ISBN 0-16-048012-4